D1530548

THE

FOLD-OUT ATLAS

OF THE

HUMAN BODY

A THREE-DIMENSIONAL BOOK

THE

FOLD-OUT ATLAS

OF THE

HUMAN BODY

A THREE-DIMENSIONAL BOOK

BY

ALFRED MASON AMADON, M.D.

With a new foreword by
NICHOLAS FIEBACH, M.D.
Associate Professor of Medicine
Yale University

GRAMERCY BOOKS

NEW YORK • AVENEL

Originally published in a slightly different form under the title *Atlas of Physiology and Anatomy of the Human Body.*

This 1991 edition was previously published by Bonanza Books and has been reprinted by Gramercy Books, distributed by Outlet Book Company, Inc., a Random House Company, 40 Engelhard Avenue, Avenel, New Jersey 07001.

Printed and bound in Singapore.

Library of Congress Cataloging in Publication Data

Amadon, Alfred Mason.
    The fold-out atlas of the human body: a three-dimensional book
    Originally published: Atlas of physiology and anatomy of the human body, 1906.
    1. Anatomy, Human—Atlases. I. Amadon, Alfred Mason.
Atlas of physiology and anatomy of the human body.
II. Title. [DNLM: QS A481a 1906a]
QM25.A48   1984   611   84-12426
ISBN 0-517-45127-1

10 9 8 7 6 5 4 3

CONTENTS

FOREWORD

Men and women have always been in awe of the human body. Its magnificent shapes and wondrous capabilities have inspired countless forms of creative and cultural expression. Less celebrated but no less amazing has been the revelation of what the body looks like on the inside, what its parts are, and how they work. Anatomy, the science of the internal architecture of the body, has traditionally been the province of a select few, invested with authority and viewed with suspicion. Today, technological innovation and changing societal mores have made it easier for medical scientists to study the structure and function of the body and for lay people to obtain information about it. Yet, knowledge about how the body's organs appear and where they are located remains a common mystery for most people.

Anatomy is still the first course taught in most medical schools. It is usually called *gross anatomy*—the study of the internal organs by the unaided senses of sight and touch. It is a logical beginning, learning to recognize and locate the component organs before going on to study their microscopic processes and the diseases that derange them. But it is also an initiation, a rite of passage into the fraternity of those who know what the inside of the body is like. Medical students learn gross anatomy directly, by dissecting a human cadaver and physically inspecting the inside parts. This is a powerful experience, full of strong emotions and remarkable sights.

Most people will not have the opportunity to view a dissected body and probably would not care to. But almost everyone shares the urge to know something about what goes on inside one. Natural curiosity about the bodies we inhabit makes us wonder what a muscle looks like, what is inside the nose, where food goes after it is swallowed. Everyday experiences sometimes make us pause to ask questions such as, How large is the stomach? Or, Where are the kidneys? When we are well, the smooth functioning of the body inspires an appreciation of its parts and a desire to see and understand them. When we are ill or injured, we are often impelled to know more about the part that has the problem and where it is located.

This book should provide excellent answers to those who wish to see what is there inside their own body. First published in 1906, it is beautifully drawn and superbly crafted. Little is known about the author, Dr. Alfred Mason Amadon, but, looking back on his work, we can surmise that he was an accurate and comprehensive anatomist. His drawings, which were probably based on actual human dissections, are complete and correct. The text is clear, concise, and offers straightforward descriptions of the parts of the body. There are also occasional interesting notes on the origins of terms. The special feature of the book, the intricately constructed three-dimensional fold-outs, still have great appeal. They allow the reader a sense of discovery as different levels of the body are successively displayed, and provide an appreciation of the spatial relations between the organs which is crucial to an understanding of anatomy.

While the contents of the book are surprisingly accurate, its unusual format attests to the historical origin of the current edition. For example, special sections with closer views of the pharynx and larynx, tooth, stomach, ear, nose, and eye are provided. This extra emphasis on the organs of digestion and the special senses probably reflects the fact that these were the parts of the body that were most accessible and of greatest interest to medical scientists at the turn of the century. Regrettably, there are no descriptions of female anatomy.

For the serious student of anatomy or the merely curious, this book will unfold many of the mysteries of the human body. It will teach us something about ourselves and may make us better patients—and better doctors. Perhaps this reminder of the remarkable structure of the body will make us more appreciative of its complexity, more careful with its maintenance, and more sanguine about its prospects. At the very least, it will give us another occasion to wonder about the marvelous intricacy of the human body.

New York   NICHOLAS H. FIEBACH, M.D.

PLATE I

ANATOMY OF THE HUMAN BODY

Leaf I. — THE VISCERA, RESPIRATORY, AND URINARY APPARATUS

1	Trachea	The windpipe. The name *trachea* means "rough," and is given because of the firm structure of the windpipe. This organ is a membranous tube held open throughout its length by a close series of cartilaginous rings which are incomplete behind and are set into the membranous walls. Through it and its branches (bronchi and bronchioles) air is conveyed to and from the lungs.
2	Apex of the Right Lung	The lungs, heart, great blood-vessels, and œsophagus fill the cavity of the thorax, or chest. The bases of the lungs rest upon the diaphragm. (See No. 31.)
3	Apex of the Left Lung	
4	Base of the Right Lung	
5	Base of the Left Lung	
6	Ramifications of the Pulmonary Artery	Through the pulmonary artery vitiated (venous) blood is taken from the right ventricle of the heart to the lungs, and by the ramifications of the artery is distributed throughout the delicate lung tissue. The terminal branches of the artery are very small and have very thin walls. In these delicate terminals the venous blood is relieved of its carbon dioxide gas and recharged with oxygen from the air that is in the lungs. Thence, purified and refreshed, the blood (now arterial) is conveyed through the ramifications of the pulmonary vein into the pulmonary vein and through it into the left auricle of the heart. (See Leaf V, No. 56.)
7	Ramifications of the Pulmonary Vein	
8	Right Ventricle of the Heart	The right side of the heart receives vitiated (venous) blood from all parts of the body except the lungs, and sends it to the lungs for purification and recharging with oxygen. The left side of the heart receives from the lungs purified and refreshed (arterial) blood, and forces it into all parts of the body except the lungs. Blood is received from veins into the auricles; by them is forced through valved openings into the thick, muscular-walled ventricles. From these it is forced through arteries into all parts of the body
9	Left Ventricle of the Heart	
10	Right Auricle of the Heart	
11	Left Auricle of the Heart	
12	Superior Vena Cava	The great vein that receives the venous blood from the whole of the upper half of the body and conveys it into the right auricle

13	Ascending Aorta	The first portion of the great artery. This artery is the main trunk of a series of vessels which convey arterial blood to all parts of the body for its nourishment. (See Leaf V, No. 69.)
14	Pulmonary Artery	The main trunk of a series of vessels that convey venous blood to the lungs for purification.
15	Inferior Vena Cava	The great vein that receives and conveys into the right auricle the venous blood from all parts of the body below the diaphragm.
16	Hepatic Veins	Three veins that convey from the liver into the inferior vena cava blood that the liver has received from the small intestines through the mesenteric veins. (See Plate III, Fig. III, No. 24.)
17	Common Iliac Veins	By their union these form the inferior vena cava.
18	Internal Iliac Veins	These receive blood from the organs that lie in the pelvis.
19	External Iliac Veins	These receive blood from the legs, and unite with the internal iliac veins to form the common iliac veins.
20	Renal Veins	These convey blood from kidneys, the left suprarenal capsule, and the left spermatic vein into the inferior vena cava.
21	Spermatic Veins	
22	Abdominal Aorta	That portion of the great arterial trunk which lies below the diaphragm. See Leaf V, No. 69.
23	Phrenic Artery	This supplies blood to the diaphragm.
24	Superior Mesenteric Artery	This lies in the mesentery, which is the membranous sling that suspends the intestines from the posterior wall of the abdomen. This artery supplies blood to all the small intestines except the first part of the duodenum, and to the upper half of the large intestine.
25	Renal Arteries	These supply blood to the kidneys and to the suprarenal capsules.
26	Spermatic Arteries	
27	Inferior Mesenteric Artery	This supplies blood to the lower half of the large intestine.
28	Common Iliac Arteries	The two branches in which the abdominal aorta terminates. They take their name from the ilium bone, against the inner surface of which they lie.
29	Internal Iliac Artery	The branch of the common iliac artery that supplies blood to most of the organs contained in the pelvis.
30	External Iliac Artery	The branch of the common iliac artery that supplies blood to the legs and feet.
31	Diaphragm	The midriff. The muscular partition dividing the chest cavity from that of the abdomen. It is the chief muscle of breathing.

32	Right Kidney	The kidneys take from the blood certain waste products
33	Left Kidney	and water, which together constitute urine.
34	Right Suprarenal Capsule	The suprarenal capsules are concerned with the tone of the walls of the arteries, thus regulating arterial blood
35	Left Suprarenal Capsule	pressure.
36	Right Ureter	The ureters conduct the urine from the kidneys to the
37	Left Ureter	bladder.
38	Urinary Bladder	A reservoir for the storage of urine until it is excreted from the body.

LEAF II. — THE MUSCLES

NOTE. A *flexor* bends a joint. An *extensor* straightens a joint. A *supinator* turns the palm upward. A *pronator* turns the palm downward. An *adductor* draws a limb toward the middle line of the body. An *abductor* moves a limb from the middle line of the body.

Muscles are named from their situation, from their use, from their shape, from their divisions, from their direction, or from the points of their attachment.

1	Occipito-frontalis (Frontal Portion)	The forehead muscle, wrinkling the skin of the forehead.
2	Temporalis	The temple muscle — a strong chewing muscle, its tendon being attached to the lower jaw. (See Plate II, Leaf II, No. 13.)
3	Orbicularis Palpebrarum	Circular muscle of the eyelids. Closes eyes. (See Plate II, Leaf II, Nos. 2 and 3.)
4	Zygomaticus Minor	These raise the upper lip and the corners of the mouth.
5	Zygomaticus Major	
6	Buccinator	The muscle of the cheek. Trumpeter's muscle.
7	Sterno-cleido-mastoideus	The prominent round muscle of the neck. It rotates the head and flexes the head upon the neck. (See Plate II, Leaf II, Nos. 23–25.)
8	Platysma Moyides	A thin, flat muscle that stretches the skin of the neck.
9	Levator Labii Superioris	Lifter of the upper lip.
10	Levator Labii Superioris et Alae Nasi	Lifter of the upper lip and of the sides of the nose.
11	Orbicularis Oris	The circular muscle of the mouth. It closes the lips and by forcible action purses the lips.
12	Quadratus Menti or Depressor Labii Inferioris	Square muscle of the chin. Depresses the lower lip.
13	Sterno-cleido-mastoideus	(See No. 7.)
14	Scalenus Anticus, Medius et Posticus	Oblique muscles of the neck. They bend the neck.
15	Trachea	(See Leaf I, No. 1.)
16	Pectoralis Minor	The smaller breast muscle, lying beneath the great breast muscle. It is attached to the shoulder-blade and depresses it.

17	Intercostales	Short muscles stretched between the successive ribs. They move the ribs in breathing.
18	Pectoralis Major	The great breast muscle. It is attached to the upper end of the humerus and abducts the arm. (See No. 42.)
19	Serratus Magnus	The great serrated (notched) muscle. It is attached to the shoulder-blade and braces it in pushing.
20	Subclavius	This passes from the first rib to the under surface of the collar-bone. It depresses the collar-bone.
21	Obliquus Externus Abdominis	The external oblique muscle of the abdomen. This and the two following muscles compress and support the contents of the abdomen.
22	Rectus Abdominis	(See No. 21.)
23	Aponeurosis of Obliquus Externus Abdominis	The flat tendon of the external oblique muscle.
24	Umbilicus	The navel.
25	Obliquus Internus Abdominis	(See No. 21.)
26	Deltoid	The muscle of the shoulder, shaped like the Greek letter *delta*. It raises the arm from the side.
27	Biceps Flexor Cubiti	The anterior muscle of the arm. Flexor of the forearm. (See Nos. 43 and 44.)
28	Triceps Extensor Cubiti	The posterior muscle of the arm. Extensor of the forearm.
29	Pronator Radii Teres	The round pronator of the forearm.
30	Supinator Longus	The long supinator of the forearm.
31	Flexor Carpi Radialis	The flexor of the wrist.
32	Palmaris Longus	This tightens the skin of the palm.
33	Anterior Annular Ligament of the Wrist	A fibrous bracelet under which pass the tendons of the forearm muscles which flex the fingers.
34	Extensor Brevis Pollicis	The short extensor of the thumb.
35	Abductor Pollicis	The abductor of the thumb.
36	Tendons of the Flexors of the Fingers	
37	Palmaris Brevis	This tightens the skin of the palm.
38	Coraco-brachialis	This is stretched between the coracoid process of the shoulder-blade (see Leaf III, No. 28) and the upper end of the humerus. It adducts the arm.
39	Head of the Humerus	
40	Long Tendon of the Biceps Flexor Cubiti	(See No. 27.)
41	Deltoid	(See No. 26.)

42	Tendon of the Pectoralis Major	(See No. 18.)
43	Short Head of the Biceps Flexor Cubiti	(See No. 27.) From these two heads (Nos. 43 and 44) the biceps (*two-headed*) flexor cubiti takes its name.
44	Long Head of the Biceps Flexor Cubiti	
45	Triceps Extensor Cubiti	(See No. 28.)
46	Extensor Carpi Radialis Longior	The longer extensor of the wrist.
47	Extensor Longus Pollicis	The long extensor of the thumb.
48	Extensor Communis Digitorum	The extensor of the fingers.
49	Extensor Carpi Radialis Brevior	The shorter extensor of the wrist.
50	Flexor Longus Pollicis	The long flexor of the thumb.
51	Pronator Radii Teres	The round pronator of the forearm.
52	Poupart's Ligament	A strong ligament that extends from the upper surface of the pubis bone to the crest of the ilium or hip-bone. It forms the lower border of the aponeurosis of the externus obliquus abdominis. (See No. 23.) Over its lower end and beneath it through tendinous canals pass muscles, vessels, and nerves from the abdominal and pelvic cavities. These canals furnish little resistance to pressure from within, and some of the contents of the abdominal cavity often push through them, making a hernia.
53	Inguinal Canal	For the passage of the spermatic cord and vessels. The seat of an inguinal hernia. (See No. 52.)
54	Gluteus Medius	The middle buttock muscle. It abducts the thigh.
55	Pyriformis	This holds the pelvis steady upon the femur.
56	Pectineus	Adductors of the thigh.
57	Adductor Longus	
58	Adductor Magnus	
59	Vastus Internus	These with the rectus femoris (75) form the great quadriceps (four-headed) extensor of the leg. They make the muscular mass in front and upon both sides of the femur and are inserted by one tendon (61) into the patella (62).
60	Vastus Externus	
61	Tendon of Extensors of the Leg	
62	Patella	The kneepan. Having a strong tendinous attachment to the tibia below, and to the great extensors of the leg above, this bone plays against the moving parts of the knee joint, and from its position increases the purchase of the extensors.
63	Gastrocnemius	With the soleus (87) this forms the calf of the leg. Extensor of the foot.
64	Flexor Longus Digitorum	Long extensor of the toes.

65	Tendon of Achilles	The strong tendon of the extensors of the foot.
66	Extensor Longus Hallu-cis	Long extensor of the great toe.
67	Extensor Longus Digi-torum	Long extensor of the toes.
68	Tendons of the Exten-sors of the Toes	
69	Crest of the Ilium, or Hip-bone	
70	Ilio-psoas	Flexor of the thigh upon the pelvis.
71	Tensor Vaginae Femoris	Tensor of the fibrous sheets that encase the outside muscles of the thigh (fascia lata).
72	Pectineus	
73	Adductor Longus	Adductors of the leg.
74	Vastus Externus	(See No. 60.)
75	Rectus Femoris	(See Nos. 59 and 60.) Called rectus on account of its *straight* course.
76	Sartorius	Flexor of the leg. The name means tailor's muscle, and was given to it because it was supposed to assist in crossing the legs in the squatting position. It is the longest muscle in the body.
77	Vastus internus	(See No. 59.)
78	Peroneus Longus et Brevis	Extensors of the foot. The name is derived from the Latin word meaning " boot."
79	Tibialis Anticus	Flexor of the foot.
80	Tibia	
81	Extensor Longus Digi-torum	The long extensor of the toes.
82	Extensor Proprius Hal-lucis	The extensor of the great toe.
83	Anterior Annular Liga-ment of the Ankle	A fibrous anklet under which pass the tendons of the extensor muscles.
84	Extensor Brevis Digi-torum	The short extensor of the toes.
85	Extensor Brevis Hal-lucis	The short extensor of the great toe.
86	Gastrocnemius	
87	Soleus	Extensors of the foot. (See No. 63.)

Leaf III. — THE SKELETON

(Anterior View)

1	Frontal Bone	The forehead bone, forming also portions of the floor of the cranium (see note after No. 4), and of the orbit (see No. 8).
2	Parietal Bones	These form the outer walls of the cranium (hence the name).
3	Temporal Bones	These form part of the outer wall and part of the floor of the cranium. They contain the delicate and complicated hearing apparatus.
4	Sphenoid Bone	This forms much of the floor of the cranium and part of the outer walls. The name is from a word meaning wedge.
		NOTE. These bones, with the occipital (Leaf IV, No. 1) and ethmoid (see Plate II, Leaf IV, No. 14) bones, form the cranium or that part of the skull containing the brain. (See Plate II, Leaf IV, No. 15.)
5	Zygoma	The cheek-bone, formed partly by a ridge of the malar bone and partly by one from the temporal bone. The name is from a Greek word meaning " yoke."
6	Mastoid portion of the Temporal Bone	A thick mass, spongy within, the cells communicating with the middle ear. (See Plate II, Leaf V, No. 19.)
7	Nasal Bones	These form the bridge of the nose.
8	The Orbits	Bony recesses for the eyes. (See Plate IV, Fig. III, Nos. 58–61.)
9	Superior Maxillary Bones	These form part of the roof of the mouth and floor of the nose, and also support the teeth. Their interior is hollow, the cavity being called the antrum of Highmore. (See Plate IV, Fig. II, No. 22.)
10	Nasal Cavities	
11	Teeth	
12	Inferior Maxillary Bone	(See Plate II, Leaf V, No. 9.)
13	7th Cervical Vertebra	The upper seven vertebrae are in the neck, and on this account are called cervical (neck) vertebrae.
14	1st Dorsal Vertebra	The twelve vertebrae to which ribs are attached are called dorsal (back) vertebrae.
15	12th Dorsal Vertebra	
16 to 20	Lumbar Vertebrae	The name comes from a Latin word meaning loins.
21	Sacrum	This is made of five vertebrae consolidated.
22	Coccyx	This is made of four vertebrae. The name means cuckoo, from the bone's similarity to the beak of that bird.

23 24 25	Manubrium Gladiolus Xiphoid Appendix } Sternum	The breastbone, to which the ribs are attached in front.
26	Clavicle	The collar-bone.
27	Scapula	The shoulder-blade. The name is from a Latin word meaning a spade. NOTE. The clavicles and scapulae form a shoulder girdle about the top of the thorax and furnish a well-braced but flexible support for the arms.
28	Coracoid Process of the Scapula	
29 to 35	True Ribs	The true ribs are the upper seven. They are joined to the vertebral column behind; and in front, by means of costal cartilages, to the sternum. (See Nos. 23–25.)
36 to 40	False Ribs	The lower five ribs are false ribs. Of these the three upper are joined each to the one above by cartilage. The last two are floating ribs, attached only to the vertebral column.
41	Costal Cartilages	(See Nos. 29–35.)
42	Ilium	Hip-bone. Called ilium because it supports the flanks (ilia).
43	Promontory of the Sacrum	
44	Crest of the Ilium	The prominent portion of the hip-bone.
45	Os Pubis	
46	Ischium	The name is from a word meaning " hip." NOTE. The sacrum, ilia, ischia, and ossa pubis form one consolidated hip girdle, furnishing a rigid support for the body upon the legs and a basin-like support (the pelvis) for the organs of the abdominal cavity. The ilium, ischium, and os pubis cannot be separated and are often considered one bone, to which is given the name os innominatum.
47	Obturator Foramen, or Thyroid Foramen	An opening through the os innominatum about which its three constituent bones are assembled.
48	Shaft of the Humerus	
49	Head of the Humerus	This furnishes a round surface for a free-moving joint.
50	Radial Head of the Humerus	Against this the head of the radius plays.
51	Trochlea	The humerus portion of the joint with the ulna.
52	Ulna	The word means elbow. The ulna furnishes the main strength of the elbow joint.

53	Radius	A fibrous ring from the ulna encircles the upper end of the radius, holding it against the radial head of the humerus and also allowing it to rotate in pronation and supination of the forearm. A similar arrangement about the lower end of the ulna allows the wrist bones to rotate against it. Thus freedom of motion is secured for the hand.
54	Carpus	The carpus, or wrist, contains eight small bones arranged in two compact rows of four each.
55	Metacarpus	The bones of the palm, — five in number.
56	Phalanges	The bones of the fingers and thumb. The name is suggested by their arrangement in ranks, like a phalanx of an army.
57	Shaft of the Femur	
58	Head of the Femur	This is shaped like a ball, and fits into a socket of the os innominatum, making a ball-and-socket joint.
59	Neck of the Femur	
60	Greater Trochanter	Prominent enlargements which afford leverage to the muscles that rotate the thigh.
61	Lesser Trochanter	
62	Inner Tuberosity of the Femur	To the tuberosities are attached the strong fibrous bands that hold together the bones forming the knee joint.
63	Outer Tuberosity of the Femur	
64	Patella	The kneepan. (See Leaf II, No. 62.)
65	Tibia	The shin-bone. The word means a "flute" or "pipe."
66	Fibula	The word means a "clasp."
67	Internal Malleolus	These form the prominences on the inner and outer sides of the ankle.
68	External Malleolus	
69	Tarsus	The bones beneath the ankle joint. There are seven.
70	Metatarsus	The bones of the instep, five in number.
71	Phalanges	The bones of the toes. (See No. 56.)

Leaf IV. — THE SKELETON

(Posterior View)

1	Occipital Bone	This forms the posterior portion of the cranium.
2	Parietal Bones	(See Leaf II, No. 2.)
3	Temporal Bone	(See Leaf II, No. 3.)

4	Basilar Process of the Occipital Bone	The thick anterior portion of the occipital bone. By this portion the occipital bone rests upon the first vertebra, and through a round aperture (foramen magnum) in this portion of bone, the spinal cord enters the cranium.
5	Inferior Maxillary Bone	(See Plate II, Leaf V, No. 9.)
6	Sagittal Suture	The suture between the parietal bones. It is so called because it has the direction of an arrow's flight.
7	Lambdoidal Suture	The suture between the occipital and the parietal bones; so called because it resembles the Greek letter lambda.
8	1st Cervical Vertebra	Called the atlas, because it bears the skull.
9	2nd Cervical Vertebra	Called the axis. The anterior portion of the atlas is a bony ring which surrounds a bony pivot that rises from the axis. Upon this pivot as an axis the atlas turns. This arrangement allows free turning of the head.
10 to 14	3d to 7th Cervical Vertebrae	(See Leaf III, No. 13.)
15	1st Dorsal Vertebra	(See Leaf III, No. 14.)
16	12th Dorsal Vertebra	
17	1st Lumbar Vertebra	(See Leaf III, No. 16.)
18	5th Lumbar Vertebra	
19	Ribs	(See Leaf III, Nos. 29–40.)
20	Scapula	(See Leaf III, Nos. 26 and 27.)
21	Spine of the Scapula	A thick bony ridge for attachment of muscles. It ends in the acromion process. (See No. 23.)
22	Clavicle	(See Leaf III, No. 26.)
23	Acromion Process of the Scapula	The bone surmounting the shoulder-joint.
24	Shaft of the Humerus	(See Leaf III.)
25	Head of the Humerus	
26	Shaft of the Ulna	
27	Olecranon Process of the Ulna	A stout bony process to which is attached the triceps extensor of the forearm. Along the inner border of the olecranon passes the ulnar nerve, a blow upon which causes numbness along the ulnar side of forearm. For this reason the olecranon is called the " crazy bone."
28	Radius	
29	Carpus	
30	Metacarpus	
31	Phalanges	
32	Sacrum	(See Leaf III.)
33	Coccyx	
34	Ilium	
35	Os Pubis	
36	Ischium	

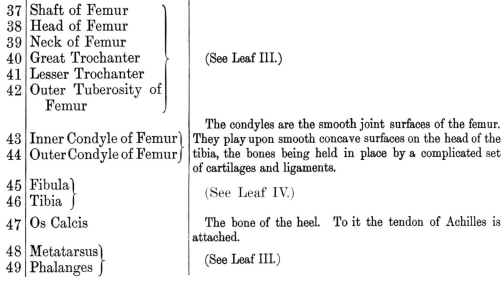

37	Shaft of Femur	
38	Head of Femur	
39	Neck of Femur	
40	Great Trochanter	(See Leaf III.)
41	Lesser Trochanter	
42	Outer Tuberosity of Femur	

		The condyles are the smooth joint surfaces of the femur. They play upon smooth concave surfaces on the head of the tibia, the bones being held in place by a complicated set of cartilages and ligaments.
43	Inner Condyle of Femur	
44	Outer Condyle of Femur	
45	Fibula	(See Leaf IV.)
46	Tibia	
47	Os Calcis	The bone of the heel. To it the tendon of Achilles is attached.
48	Metatarsus	(See Leaf III.)
49	Phalanges	

LEAF V. — VISCERA, BLOOD-VESSELS, AND NERVES

1	Facial Artery	
2	Temporal Artery	
3	Facial Vein	
4	Temporal Vein	
5	Frontal and Supraorbital Veins	(For vessels and nerves of the face, see Plate II.)
6	Supraorbital Nerve	
7	Temporal Nerve	
8	Malar and Infraorbital Nerves	
9	Maxillary Nerve	
10	Mental Nerve	

11	Cervical Plexus of Nerves	Through openings between each two adjacent vertebrae pass a pair of nerves, having their origin in the spinal cord. Upon its emergence from its bony canal each nerve divides into an anterior and a posterior branch. The posterior branches supply the neighboring muscles and skin. The anterior branches in the cervical, lumbar, and sacral regions, by uniting with each other, form a net or tangle of nerve fibres called a plexus, and from the plexus nerves pass to the neighboring muscles and skin. The anterior divisions of the first four cervical nerves form the cervical plexus. The anterior divisions of the lower four cervical nerves and part of the anterior division of the first dorsal nerve form the brachial plexus. The nerves from the cervical plexus supply many of the neck muscles and the skin. Those from the brachial plexus supply some of the neck and back muscles, but mostly the muscles and skin of the arm.
	NOTE. — Nerves that supply the skin render it sensitive; nerves that supply muscles control the action of those muscles.	

12	Spinal Cord	The great nerve trunk, arising from the brain and extending downward through a bony canal formed by bony arches behind the bodies of the vertebrae. (See Plate II, Leaf III, No. 9.) The spinal cord terminates at the level of the lower surface of the first lumbar vertebra.
13	Intercostal Nerves	These supply the muscles and skin of the back, chest, and abdomen.
14	Ganglia of Sympathetic Nervous System	The sympathetic nervous system consists of a series of ganglia, or nodules, connected together by intervening nerve threads and extending from the base of the skull to the coccyx, one series on each side of the middle line of the body and in front of or on each side of the vertebral column. Connected with this series are three great gangliated plexuses, — the cardiac plexus in the thorax, the solar or epigastric plexus, behind the stomach, and the hypogastric plexus, between the common iliac arteries. (A plexus is a network ; see No. 11.) Fibres from this system communicate with the cerebro-spinal nerves, and other fibres supply the viscera and walls of the blood-vessels. The nerves of the sympathetic system control the muscular walls of the viscera and vessels, and regulate the action of glands.
		NOTE. There is no real separation of the nervous system into two parts — the cerebro-spinal system and the sympathetic system. The two are intimately connected and are parts of one system. Merely for convenience of description they are treated separately.
15	12th Dorsal Nerve	
16	1st Lumbar Nerve	
17	5th Lumbar Nerve	
18	Lumbar Plexus of Nerves	(See No. 11.) This gives branches to the skin and muscles of the thigh.
19	Crural Nerves	These supply the muscles of the front of the thigh (except the tensor vaginae femoris), the skin of the front and inner surfaces of the thigh, and the skin of the leg and foot.
20 to 21	Abdominal Branches of the Lumbar Nerves	
22	External Cutaneous Nerve	This passes from the lumbar plexus and supplies the skin of the front and outer sides of the thigh.
23	Sacral Plexus of Nerves	(See No. 11.) This gives branches to the muscles and skin of the buttock, thigh, leg, and foot.
24	Intercostal Nerves and Veins	
25	Cephalic Vein	
26	Anterior Ulnar Vein	
27	Median Basilic Vein	The vein usually opened by surgeons in bleeding.
28	Radial Vein	

29	Acromial Nerve	This is a branch of the cervical plexus. It supplies sensation to the skin of the shoulder.
30	Circumflex Nerve	This is a branch of the brachial plexus, supplying some of the muscles and skin of the shoulder and the shoulder-joint.
31	Internal Cutaneous and Musculo-cutaneous Nerves	This is a branch of the brachial plexus, supplying sensation to the skin of the forearm and motion to some of the muscles of the arm.
32	Basilic Vein	This, with the cephalic vein, drains the blood from the arm and hand.
33	Brachial Artery	The great artery of the arm.
34	Radial Artery	The pulse is usually felt where this artery lies near the surface, on the thumb side of the wrist.
35	Ulnar Artery	
36	Median Nerve	This nerve receives its name from the course it takes along the middle of the arm and forearm. It supplies nearly all the muscles of the forearm and of the thumb side of the hand, and also supplies the skin of the palm.
37	Musculo-spiral Nerve	This supplies the muscles and skin of tne back of the arm and forearm, and the skin of the back of the hand.
38	Superficial Palmar Arch	An arterial arch across the palm, made by the ulnar artery and completed by a small branch from the radial artery. From this arch arteries branch downwards to supply the fingers. The radial artery forms similarly a deeper arch in the palm.
39	Femoral Artery	This passes out of the abdomen under Poupart's ligament. (See Leaf II, No. 52.) It is the great artery of the leg.
40	Femoral Vein	
41	Internal Saphenous Vein	A long vein extending along the inner surface of the leg from the foot to the groin. It lies near the surface and its walls often become dilated from the weight of the blood. Veins thus dilated are called varicose veins.
42	Middle Cutaneous Nerve	This supplies sensation to part of the skin of the thigh.
43	Internal Cutaneous Nerve	This supplies sensation to the skin of the inner surface of the leg.
44	Internal Saphenous Nerve	This supplies sensation to the skin about the patella and over the front of the leg.
45	Musculo-cutaneous Nerve	This supplies motion to the muscles on the outer side of the leg and sensation to the skin of the top of the foot.
46	Femoral Artery	
47	Deep Femoral Artery	This supplies blood to the muscles of the back of the thigh.
48	Anterior Tibial Artery	This supplies the anterior muscles of the leg and knee.
49	Anterior Tibial Nerve	This supplies the extensor muscles of the foot.
50	Dorsalis Pedis Artery	This supplies the top of the foot and the toes.

51	Larynx	(See Plate III, Fig. I.)
52	Thyroid Gland	This is concerned with the elaboration of the blood.
53	Trachea	(See Leaf I, No. 1.)
54	Right Bronchus	The bronchi and bronchioles (see No. 60) are also
55	Left Bronchus	known as the bronchial tubes.
56	Right Lung	The ultimate divisions of the bronchial tubes are tiny
57	Left Lung	and thin-walled, and toward their ends they become irregularly and increasingly dilated, ending in a mass of small air-cells. These have elastic walls and are lined with thin mucous membrane, richly supplied with capillaries. The mass of these air-cells is the lung. The interchange of oxygen and carbon dioxide between the air and the blood takes place in the walls of the cells. (See Leaf I, No. 6.)
58	Branches of Pulmonary Artery	
59	Branches of Pulmonary Vein	(See Leaf I, Nos. 6 and 7.)
60	Bronchioles	The small divisions of the bronchi. (See Nos. 54 and 55.)
61	Heart	
62	Left Ventricle of the Heart	
63	Right Ventricle of the Heart	
64	Right Auricle of the Heart	(See Leaf I, Nos. 8–11.)
65	Left Auricle of the Heart	
66	Superior Vena Cava	(See Leaf I, No. 12.)
67	Right Innominate or Brachio-cephalic Vein	These vessels receive the venous blood from the head, arms, and upper part of the chest, and unite to form the
68	Left Innominate or Brachio-cephalic Vein	superior vena cava.
69	Aorta	The great arterial trunk. It rises from the left ventricle of the heart, arches backward and to the left in the two of the chest cavity, and follows a downward course, lying somewhat to the left of the vertebral column. It terminates at the fourth lumbar vertebra.
70	Pulmonary Artery	
71	Pulmonary Vein	(See Leaf I, Nos. 6 and 7.)
72	Inferior Vena Cava	(See Leaf I, No. 15.)
73	Diaphragm	(See Leaf I, No. 31.)
74	Oesophagus	
75	Stomach	
76	Inner Surface of the Stomach	(See Plate III, Fig. III.)

77	Cardiac Orifice of the Stomach	
78	Pyloric Orifice of the Stomach	(See Plate III, Fig. III.)
79	Spleen	
80	Pancreas	
81	Duodenum	
82	Ileum	The lower three-fifths of the small intestine, opening below into the large intestine.
83	Caecum and Vermiform Appendix	The caecum is the pouch-like upper end of the large intestine. The vermiform appendix is the shrivelled remnant of the lengthened caecum found in many mammals.
84	Ascending Colon	
85	Transverse Colon	
86	Descending Colon	Divisions of the large intestine.
87	Sigmoid Flexure	
88	Rectum	
89	Liver	
90	Section of the Liver	
91	Under Surface of the Liver	(See Plate III.)
92	Gall Bladder	
93	Right Kidney	(See Leaf I, Nos. 32 and 33.)
94	Left Kidney	
95	Ureters	(See Leaf I, Nos. 36 and 37.)
96	Psoas Muscle	
97	Urinary Bladder	(See Leaf I, No. 38.)
98	Orifices of Ureters	
99	Inner Surface of the Bladder	
100	Orifice of the Urethra	

PLATE II
THE HEAD AND NECK

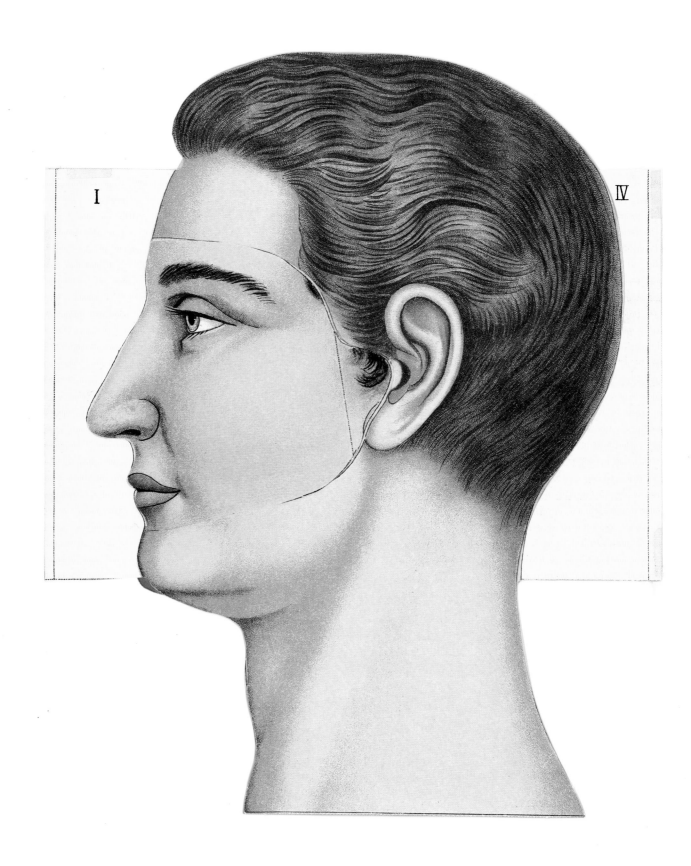

PLATE II

THE HEAD AND NECK

Leaf I.—SKELETON OF THE FACE

1	Anterior Floor of the Cranial Cavity	The cranium is the bony case containing the brain. (See Plate I, Leaf III, No. 4, note.)
2	Frontal Sinus	(See Plate IV, Fig. II.)
3	Frontal Bone	(See Plate I, Leaf III, No. 1.)
4	Nasal Bone	
5	Sphenoidal Sinus	
6	Superior Turbinated Body	
7	Middle Turbinated Body	
8	Inferior Turbinated Body	(See Plate IV, Fig. II.)
9	Superior Meatus	
10	Middle Meatus	
11	Inferior Meatus	
12	Vestibule of the Nasal Cavity	
13	Mouth of the Eustachian Tube	This communicates with the middle ear. (See Plate IV, Fig. I, No. 21.)
14	Upper Lip	
15	Superior Maxillary Bone	(See Plate I, Leaf III, No. 9.)
16	Soft Palate	This forms the posterior portion of the roof of the mouth. It is a muscular curtain which, in the act of swallowing, falls back against the posterior wall of the pharynx and so prevents food from being forced into the back of the nose.
17	Hard Palate	This forms the anterior part of the roof of the mouth. (See Plate III, Fig. I, No. 46.)
18	Molar Teeth	The grinding teeth, three on each side of each jaw and the farthest back.
19	Incisor Teeth	Cutting or gnawing teeth, two on each side of each jaw, nearest the middle line.
		NOTE. The third tooth from the middle line in each jaw is the canine or eye tooth — the tearing tooth of carnivorous animals. The fourth and fifth from the middle line in each jaw are the pre-molars, or bicuspids.

20	Lower Lip	
21	Inferior Maxillary Bone	(See Leaf V, No. 9.)
22	Zygoma	(See Plate I, Leaf III, No. 5.)
A	The Eyeball	
b	Pupil	
c	Iris	
d	Sclera	
e	Superior Rectus Muscle	
f	External Rectus Muscle	
g	Inferior Rectus Muscle	
h	Superior Oblique Muscle	
i	Inferior Oblique Muscle	
j	Optic Nerve	
k	Retina	(See Plate IV, Fig. III.)
l	Choroid Coat of the Eye	
m	Cavity occupied by the Vitreous Humor	
n	Crystalline Lens	
o	Anterior Chamber, occupied by the Aqueous Humor	
p	Cornea	
q	Ciliary Body	

LEAF II. — MUSCLES, VESSELS, AND NERVES

(Nerves supplying skin render the skin sensitive. Nerves supplying muscles control the motion of those muscles.)

1	Occipito-frontalis (Frontal Portion)	This wrinkles the forehead and raises the eyebrows.
2	Orbicularis Palpebrarum (Orbicular Portion)	A circular muscle whose outer or orbicular portion is thicker than the central or palpebral (lid) portion. The palpebral portion closes the lids gently, as in sleep and in winking. The orbicular portion closes the eyes tightly with much wrinkling of the surrounding skin.
3	Orbicularis Palpebrarum (Palpebral Portion)	
4	Compressor Naris	This depresses the sides (alae) of the nose.
5	Levator Labii Superioris et Alae Nasi	This elevates the upper lip and the sides of the nose.
6	Orbicularis Oris	The circular muscle of the mouth. It closes the lips, and by forced contraction purses the lips.
7	Levator Labii Superioris	This elevates the upper lip.
8	Levator Anguli Oris	This elevates the angle of the mouth.
9	Zygomaticus Minor	These elevate the upper lip.
10	Zygomaticus Major	

11	Buccinator (Trumpeter)	The muscle lying against the inside of the cheek. The trumpeter's muscle.
12	Risorius	This draws the corner of the mouth outward, as in laughter, — hence the name.
13	Masseter	This muscle, with the temporalis (see Plate I, Leaf II, No. 2), raises the lower jaw against the upper jaw. These are the chewing muscles. (See Leaf V No. 9.)
14	Depressor Anguli Oris	This depresses the corner of the mouth.
15	Depressor Labii Inferioris	This depresses the lower lip.
16	Levator Menti	This protrudes the lower lip.
17	Attollens aurem	These are rudimentary muscles and usually have no action. At best their action is but slight in man.
18	Attrahens aurem	They (17) raise, (18) draw forward, and (19) draw backward the ear.
19	Retrahens aurem	
20	Occipito-frontalis (Occipital Portion)	This draws the scalp backward. It is usually powerless.
21	Aponeurosis of the Occipito-frontalis	A tendinous sheet connecting the two portions of the occipito-frontalis muscle.
22	Trapezius	A large, flat, triangular muscle attached to the backbone from the occipital bone down to the last dorsal vertebra, its fibres converging to the junction of the collar-bone and shoulder-blade. It draws the shoulder back.
23	Sterno-cleido-mastoideus (Mastoid Portion)	
24	Sterno-cleido-mastoideus (Sternal Portion)	The prominent round muscle that extends from the mastoid portion of the temporal bone to the sternum (breast-bone) and clavicle (collar-bone). It turns the head toward the opposite side and inclines it toward the shoulder of the same side, in which action it stands out prominently.
25	Sterno-cleido-mastoideus (Clavicular Portion)	
26	Splenius Capitis	These muscles pass upward from the posterior surfaces of the upper dorsal and lower cervical vertebrae to the occipital bone and the upper cervical vertebrae.
27	Splenius Colli	They pull the head backward.
28	Levator Anguli Scapulæ	This muscle elevates the scapula, as in shrugging the shoulders.
29	Scaleni (Anticus et Posticus)	These muscles draw the neck forward or to one side.
30	Digastric	A double muscle, its two parts united, end to end, by a tendinous centre which is held by a loop to the side of the hyoid bone. The ends of the muscle are attached to the mastoid portion of the temporal bone and to the lower border of the lower jaw. Thus it suspends and raises the hyoid bone.

31	Sterno-hyoid Muscle	This depresses the hyoid bone.
32	Omo-hyoid Muscle	(See No. 62.)
33	Thyro-hyoid Muscle	This draws the hyoid bone and the thyroid cartilage together and depresses the hyoid bone.
34	Clavicle	(See Plate I, Leaf III, No. 26.)
35	Zygoma	(See Plate I, Leaf III, No. 5.)
36	Inferior Maxillary Bone	
37	Subclavian Artery	From the arch of the aorta (Plate I, Leaf V, No. 69) two large arteries are given off on the left side, — the carotid artery, which supplies the head, and the subclavian artery, which supplies the shoulder and arm. On the right side one large artery, the innominate artery, is given off from the aorta, and this divides into the right carotid artery and the right subclavian artery. The subclavian artery arches outward under cover of the clavicle and passes through the axilla (armpit) — where it is called the axillary artery — into the arm, where it is called the brachial artery.
38	Common Carotid Artery	(See No. 37.) This lies under the anterior border of the sternocleidomastoid muscle, in the same sheath with the internal jugular vein and the pneumogastric (tenth cranial) nerve. (See Leaf VI, No. 10.) It is the great artery of the neck, and its pulsations can often be felt.
39	Internal Carotid Artery	One of the two branches of the common carotid artery. It supplies the anterior part of the brain, the eye, and to some extent the forehead and nose.
40	External Carotid Artery	One of the two branches of the common carotid artery. Its branches supply most of the face, the nose, the mouth, the pharynx and part of the larynx, some of the back of the head, and some of the dura mater.
41	Facial Artery	A branch of the external carotid artery. It supplies the muscles about the lower jaw and lips and the lower part of the nose.
42	Submental Artery	A branch of the facial artery which supplies the muscles lying beneath the jaw and those of the chin.
43	Inferior Labial Artery	A branch of the facial artery supplying the muscles and skin of the lower lip.
44	Angular Artery	The termination of the facial artery, lying upon the side of the nose and ending at the inner angle of the eye. It supplies the muscles and skin of the region.
45	Supraorbital Artery	One of the terminal branches of the internal carotid artery, passing out from the orbit of the eye through an opening in the bone and passing up over the forehead, which it supplies with blood.
46	Superficial Temporal Artery	One of the two terminal branches of the external carotid artery. It passes near the surface, just in front of the ear, and as it crosses the zygoma furnishes a convenient place for feeling the arterial pulse.

47	Anterior Temporal Artery	Branches of the superficial temporal artery, supplying blood to the scalp of the side of the head.
48	Posterior Temporal Artery	
49	Posterior Auricular Artery	A branch of the external carotid artery, passing behind the ear and supplying with blood the auricle and the bone and scalp behind the ear.
50	Occipital Artery	A branch of the external carotid artery, passing back among the muscles of the back of the neck, which it supplies, and ending in the scalp of the back of the head, which it also supplies.
51	Inferior Maxillary Bone	(See Leaf V, No. 9.)
52	Hyoid Bone	(See Plate III, Fig. I.)
53	Thyroid Gland	(See Plate I, Leaf V, No. 52.)
54	Trachea	(See Plate I, Leaf I, No. 1.)
55	Clavicle	(See Plate I, Leaf III, No. 26.)
56	Parotid Gland	(See Leaf III, No. 20.)
57	Masseter	(See No. 13.)
58	Stylo-hyoid Muscle	This passes from the styloid process of the temporal bone (see Plate III, Fig. I, No. 62) to the hyoid bone, which it elevates and draws backward in the act of swallowing.
59	Hyoglossus	One of the tongue muscles arising from the hyoid bone.
60	Mylo-hyoid Muscle	This forms the muscular floor of the mouth cavity. It takes part in the first movement of swallowing.
61	Genio-hyoid Muscle	This assists the mylo-hyoid muscle in swallowing.
62	Omo-hyoid Muscle	A double muscle. The anterior portion is attached to the hyoid bone, and the posterior portion to the upper border of the shoulder-blade. These two portions of the muscle are joined by a central tendon, which passes through a fibrous loop that is fastened to the cartilage of the first rib. This muscle depresses the hyoid bone.
63	Sterno-thyroid Muscle	This passes from the sternum (breastbone) to the thyroid cartilage of the larynx. It depresses the larynx.
64	Sterno-hyoid Muscle	This passes from the sternum (breastbone) to the hyoid bone. It depresses the hyoid bone.
65	Thyro-hyoid Muscle	This is practically a continuation of the sterno-thyroid. (See No. 63.) It depresses the hyoid bone. (See No. 33.)
66	Longus Colli	This passes from the first cervical vertebra to the third dorsal vertebra. It bends the neck.
67	Scalenus Anticus	These pass from the first and second ribs to the sides of the lower six cervical vertebrae. They draw the neck forward or to one side.
68	Scalenus Medius	
69	Scalenus Posticus	
70	Levator Anguli Scapulæ	(See No. 28.)

71	Splenius Capitis et Colli	(See Nos. 26 and 27.)
72	Trapezius	(See No. 22.)
73	Sternocleidomastoideus	(See No. 23.)
74	Subclavian Artery	(See No. 37.)
75	Parotid Gland	(See Leaf III, No. 20.)
76	Submaxillary Gland	This is situated under the angle of the jaw. It furnishes part of the saliva, and its duct, called Wharton's duct, opens into the mouth at the summit of a small papilla situated on the floor of the mouth under the tongue and near the middle line.
77	Greater Occipital Nerve	The posterior branch of the second cervical nerve. (See Plate I, Leaf V, No. 11.) This supplies motion to some of the muscles of the back of the neck, the scalp over the occipital bone, and the skin of the back of the auricle.
78	Lesser Occipital Nerve	This is a branch of the cervical plexus (see Plate I, Leaf V, No. 11), supplying the skin of the scalp behind the ear and part of the skin of the auricle and the occipito-frontalis muscle.
79	Great Auricular Nerve	This passes from the cervical plexus (see Plate I, Leaf V, No. 11), supplying the skin over the angle of the jaw and in front of the ear, the skin of the back of the auricle, and the skin over the mastoid portion of the temporal bone.
80	Auriculo-temporal Nerve	A branch from the third division of the fifth cranial nerve. (See Leaf VI.) It supplies the skin of the front of the auricle and of the external auditory meatus (see Plate IV, Fig. I, No. 2), the parotid gland, and the skin of the temporal region.
81	Spinal Accessory Nerve	The eleventh cranial nerve. (See Leaf VI.) It supplies motion to some of the outer muscles of the neck.
82	3d and 4th Cervical Nerves	A portion of the cervical plexus. (See Plate I, Leaf V, No. 11.)
83	Clavicular, Sternal, and Acromial Nerves	These are branches of the cervical plexus (see Plate I, Leaf V, No. 11), supplying the skin over the shoulder and lower part of the neck anteriorly.
84	Communicans Hypoglossi	This nerve is a branch of the cervical plexus (see Plate I, Leaf V, No. 11), and communicates with the twelfth cranial nerve. (See Leaf V.)
85	Superficial Cervical Nerve	This is a branch of the cervical plexus (see Plate I, Leaf V, No. 11), supplying the skin of the front of the neck and the platysma myoides muscle.
86	Superficial Cervical Nerve (ascending branch)	This is a branch of the cervical plexus (see Plate I, Leaf V, No. 11), forming a small plexus with branches from the seventh cranial nerve. (See Leaf VI.)
87	Supramaxillary Nerves	These pass from the seventh cranial nerve (see Leaf VI), supplying the muscles of the lower lip and chin.

88	Internal Jugular Vein	This receives all the venous blood from the brain (see Leaf IV, No. 58), and from the face and neck; it follows the internal carotid artery down the neck, under the sterno-cleidomastoid muscle, and unites with the subclavian vein to form the innominate vein.
89	External Jugular Vein	This receives most of the blood from the scalp and from the deep parts of the face. It lies near the surface and empties into the subclavian vein.
90	Anterior Jugular Vein	This receives blood from the under parts of the chin and jaw and empties into the external jugular vein.
91	Facial Vein	This receives blood from the face, accompanying the facial artery, unites with a branch from the external jugular, and empties into the internal jugular beneath the sterno-cleidomastoid muscle.

LEAF III. — BLOOD-VESSELS AND NERVES

1	Subclavian Artery	(See Leaf II, No. 37.)
2	Vertebral Artery	The first branch from the subclavian artery (No.1), passing backward to enter an opening in the transverse process (see No. 9) of the sixth cervical vertebra. It ascends through similar openings (foramina) in all the cervical vertebrae above the sixth, through the foramen magnum (see Plate I, Leaf IV, No. 4), and passes forward and upward in front of the medulla oblongata. At the lower border of the pons Varolii it unites with the vertebral artery of the opposite side to form the basilar artery. (See Leaf VI, Nos. 21, 22, 25, and 26.) Its branches supply some of the muscles of the back of the neck and part of the cerebellum and dura mater.
3	Common Carotid Artery	(See Leaf II, No. 38.)
4	Thyroid Axis	A short arterial trunk, arising from the subclavian artery and dividing into three branches, which supply the larynx and upper part of the trachea, the muscles of the side of the neck, the lower hyoid muscles, and the muscles lying above the scapula.
5	Superior Intercostal Artery	A branch of the subclavian artery; it supplies some of the deep neck muscles and those of the first intercostal (between the ribs) space.
6	Deep Cervical Artery	A branch of the superior intercostal artery (No. 5) which supplies blood to some of the deep neck muscles.
7	Superior Thyroid Artery	The first branch of the external carotid artery (see Leaf II, No. 40), supplying muscles connected with the hyoid bone, the sternocleidomastoid muscle, and the upper part of the larynx.
8	First Rib	
9	Transverse Processes of the Cervical Vertebrae	Each vertebra consists of a solid disc of bone, called the body, from which an arch of bone springs behind.

		The series of arches form the canal for the spinal cord. From the sides of the bodies or of the arches strong bony spines, the transverse processes, are given off, each one in the neck having an opening for the transmission of the vertebral artery. At the summit of the arch spinous processes of bone project backward and the tips of these processes can be seen and felt in the middle line of the back.
10	Trachea	(See Plate I, Leaf I, No. 1.)
11	Thyroid Gland	(See Plate I, Leaf V, No. 52.)
12	Thyroid Cartilage	(See Plate III, Fig. I.)
13	Oesophagus	(See Plate III, Fig. III, No. 1.)
14	Frontal Vein and Supra-trochlear Nerve	The beginning of the facial vein (Leaf II, No. 91) and a terminal branch of the upper division of the fifth cranial nerve. (See Leaf VI.)
15	Supraorbital Artery and Nerve	(See Leaf II, No. 45.) The supraorbital nerve is a companion to the supratrochlear nerve (No. 14), both being terminals of the same trunk.
16	Frontal Vein	(See No. 14.)
17	Angular Artery	(See Leaf II, No. 44.)
18	Facial Artery	(See Leaf II, No. 41.)
19	Facial Vein	(See Leaf II, No. 91.)
20	Parotid Gland	The great salivary gland, situated in front of and under the ear and embracing the ramus of the lower jaw. It lies over and between the chewing muscles and is affected in the disease called mumps. Many large vessels and nerves pass through the parotid gland.
21	Stenson's Duct	The duct from the parotid gland to the mouth, opening upon the inner surface of the cheek opposite the second molar tooth of the upper jaw.
22	Transverse Facial Artery	A branch of the superficial temporal artery, supplying the parotid gland and the masseter.
23	Superficial Temporal Artery	(See Leaf II, No. 46.)
24	Auriculo-temporal Nerve	(See Leaf II, No. 80.)
25	Great Auricular Nerve	(See Leaf II, No. 79.)
26	Lesser Occipital Nerve	(See Leaf II, No. 78.)
27	Greater Occipital Nerve	(See Leaf II, No. 77.)
28	Occipital Artery	(See Leaf II, No. 50.)
29	Occipital Vein	This drains the blood from the scalp of the occipital region and empties into the internal jugular vein; sometimes into the external jugular vein.

LEAF IV. — LONGITUDINAL SECTION THROUGH THE HEAD

1	Frontal Lobe of the Cerebrum	The brain is divided into four parts — the cerebrum, the cerebellum (see No. 55), the pons Varolii (see No. 52), and the medulla oblongata (see No. 53). By the great longitudinal fissure the cerebrum is divided into two hemispheres. (See Leaf VII.) The surface of each hemisphere is marked by deep fissures and shallow sulci, or grooves. By the deep fissures each hemisphere is more or less clearly divided into lobes, which are named according to their positions; and by shallower fissures into smaller divisions called convolutions. (See Leaf V, No. 23.)
2	Parietal Lobe of the Cerebrum	
3	Temporo-sphenoidal Lobe of the Cerebrum	
4	Occipital Lobe of the Cerebrum	
5	Inner Surface of the Frontal Bone	
6	Inner Surface of the Parietal Bone	
7	Inner Surface of the Temporal Bone	
8	Inner Surface of the Occipital Bone	
9	Blood-Vessels of the Dura Mater	The large artery is the middle meningeal artery, a branch of the internal maxillary division of the external carotid artery. (See Leaf V, No. 18.) The artery lying upon the occipital bone is a meningeal branch of the occipital artery. (See Leaf II, No. 50.)
10	Frontal Bone	
11	Parietal Bone	
12	Occipital Bone	
13	Nasal Bone	
14	Ethmoid Bone	The ethmoid bone is light and cellular in structure, and roughly cubical in shape. It enters into the formation of the floor of the cranium, the inner walls of the orbits, and the upper part of the nasal cavities. Its cells or sinuses open into the nose and are prone to catarrhal disease. (See Plate IV, Fig. II, No. 22.)
15	Sphenoid Bone	The sphenoid bone consists of a central cubical body, containing cavities called sphenoid sinuses (see 16), and lateral wings. Its shape suggests a bat with spread wings. It enters into the formation of the floor of the cranium, the upper posterior regions of the nose and pharynx, and the outer walls of the orbits and of the cranium.
16	Sphenoidal Sinus	A cavity in the body of the sphenoid bone, opening into the upper and back part of the nose. It is subject to catarrhal disease. (See Plate IV, Fig. II, No. 20.)

17	Superior Turbinated Body	
18	Middle Turbinated Body	(See Plate IV, Fig. II.)
19	Inferior Turbinated Body	
20	Mouth of the Eustachian Tube	(See Plate IV, Fig. I, No. 21.)
21	Superior Maxillary Bone	
22	Soft Palate	(See Leaf I, No. 16.)
23	Mouth	
24	Inferior Maxillary Bone	(See Leaf V, No. 9.)
25	Genio-hyoglossus	A tongue muscle whose fibres radiate from their attachment on the inner surface of the lower maxillary bone. It sometimes lacks the attachment to the hyoid bone that the name indicates.
26	Genio-hyoid Muscle	(See Leaf II, No. 61.)
27	Hyoid Bone	(See Plate III, Fig. I.)
28	Pharyngeal Tonsil	Glandular tissue situated in the upper back part of the pharyngeal vault, between the mouths of the Eustachian tubes. This tissue is prone to great enlargement and degeneration in childhood, so that it obstructs breathing and causes damage to the hearing by blocking the Eustachian tubes.
29	Epiglottis	
30	Thyro-hyoid Membrane	
31	Thyroid Cartilage	(See Plate III, Fig. I.)
32	Cricoid Cartilage	
33	Ventricle of the Larynx	
34	Trachea	(See Plate I, Leaf I, No. 1.)
35	Oesophagus	(See Plate III, Fig. III, No. 1.)
36	Body of the 1st Cervical Vertebra (Atlas)	
37	Body of the 2d Cervical Vertebra (Axis)	(See Plate I, Leaf IV, Nos. 8 and 9.)
38 to 42	Bodies of the 3d to the 7th Cervical Vertebrae	(See Leaf III, No. 9.)
43	Inner Surface of Right Hemisphere of the Cerebrum	The surface of the hemisphere that faces the deep longitudinal fissure.
44	Corpus Callosum	The isthmus of nerve fibres that connects the two hemispheres of the cerebrum.
45	Velum Interpositum	A fold of the delicate membrane (pia mater) that lies next to the brain. The margin of this fold contains a network of fine blood-vessels that is called the choroid plexus.

46	Middle Commissure	The middle one of three slender isthmuses that cross from one side of the cerebrum to the other in the third ventricle. (See No. 47.)
47	Third Ventricle of the Brain	A chamber of the under side of the brain. There are five ventricles communicating with each other. The three commissures (see No. 46) cross this ventricle.
48	Corpora Quadrigemina	A small, four-lobed mass of brain matter situated just behind the third ventricle and connected with the nerve of sight.
49	Valve of Vieussens	Part of the roof of the fourth ventricle of the brain.
50	Fourth Ventricle of the Brain	(See No. 47.)
51	Pituitary Body	A small body richly supplied with blood and lying upon the top of the sphenoid bone. The ancients supposed it to be the seat of the soul. Its use is unknown.
52	Pons Varolii	The bond of union between the various portions of the brain.
53	Medulla Oblongata	An enlargement of that portion of the spinal cord which is in the cranium.
54	Spinal Cord	The nerve trunk, made of nerve fibres passing from the brain to various parts of the body, and of nerve cells. (See Plate I, Leaf V, No. 12.)
55	Cerebellum	That part of the brain which lies in the lower back portion of the cranial cavity. The cerebellum attends to the orderly movements of muscles in such complicated acts as walking.
56	Venae Galeni	The veins of the ventricles of the brain.
57	Tentorium Cerebelli	A fold of dura mater (see Leaf VI, No. 3) that separates the cerebellum from the cerebrum.
58	Torcular Herophili	A venous channel of the dura mater begins at the anterior end of the floor of the cranial cavity and runs backward in the middle line along the roof of the cranial cavity, ending in a dilated sac, the torcular Herophili, where is received also blood from other similar channels, called sinuses. From this point the blood is taken forward and downward on either side in similar channels — the lateral sinuses — and poured into the internal jugular veins. (See Leaf II, No. 88.)

LEAF V. — THE SKULL

1	Frontal Bone	
2	Parietal Bone	(See Plate I, Leaf III.)
3	Occipital Bone	

4	Nasal Bone	The nasal bones form the bridge of the nose.
5	Lachrymal Bone	The tear duct leading from the eye to the nose lies in a groove on the surface of the lachrymal bone, giving it its name.
6	Superior Maxillary Bone	(See Plate I, Leaf III, No. 9.)
7	Malar Bone	This forms the prominent portion of the cheek-bone, the outer and lower border and wall of the orbit of the eye, and — by a horizontal prominent ridge — the anterior part of the zygoma. (See Plate I, Leaf III, No. 5.)
8	Zygoma	(See Plate I, Leaf III, No. 5.)
9 10	Inferior Maxillary Bone Ramus of the Inferior Maxillary Bone	The lower jaw bone, — a crescent-shaped bone forming the lower jaw and chin. In it are set the lower teeth. The posterior ends of the bone are turned upward to form the rami or tips of the bone, and end in smooth surfaces that fit into sockets of the temporal bone just in front of the ear, making the joint of the jaw. Just in front of the ramus is a thin but strong prong called the coronoid (crown-like) process, to which are attached the strong chewing muscles. (See Plate II, Leaf II, No. 13.)
11	Great Wing of the Sphenoid Bone	(See Leaf IV, No. 15, and Plate IV, Fig. III, No. 59.)
12	Squamous Portion of the Temporal Bone	(See Plate I, Leaf III, No. 3.) The squamous or flat portion of the temporal bone forms part of the outer wall of the cranium.
13	Mastoid Portion of the Temporal Bone	(See Plate I, Leaf III, No. 6.)
14	External Auditory Canal	(See Plate IV, Fig. I, Nos. 2 and 3.)
15	Coronal Suture	The junction of the frontal and the parietal bones, across the crown of the head.
16	Lambdoidal Suture	(See Plate I, Leaf IV, No. 7.)
17	Dura Mater	(See Leaf VI, No. 3.)
18	Middle Meningeal Artery	(See Leaf IV, No. 9.)
19	Cells in the Mastoid Portion of the Temporal Bone	(See Plate I, Leaf III, No. 6.) One of these cells, called the mastoid antrum, is larger than all the rest, and the smaller ones open into it. The antrum communicates with the middle ear by a short canal, and on this account disease of the middle ear is easily extended into the mastoid bone. (See Plate IV, Fig. I, No. 17.)
20	Ear-Drum Membrane	(See Plate IV, Fig. I, No. 6.)
21	Ear Drum Membrane with Ossicles	(See Plate IV, Fig. I, Nos. 16 and 20.)
22	Groove for Lateral Sinus	(See Leaf IV, No. 50.)

| 23 | Temporo-sphenoidal Convolutions of the Cerebrum | (See Leaf IV, Nos. 1 to 4.) |
| 24 | The Cerebellum | (See Leaf IV, No. 55.) |

LEAF VI. — THE UNDER SURFACE OF THE BRAIN

1	Occipito-frontalis	(See Plate I, Leaf II, No. 1.)
2	Frontal Bone	(See Plate I, Leaf III, No. 1.)
3	Dura Mater	A thin, tough, fibrous envelope of the brain lying upon the inner surface of the bone. Next to the brain lies a delicate membranous covering called the pia mater, and between these two envelopes is an envelope of loose structure called the arachnoid (because it resembles a spider's web). The dura mater is adherent to the bone and dips into the great longitudinal fissure between the hemispheres (see Leaf IV, No. 1) and separates their adjacent surfaces. A fold of the dura mater, called the tentorium cerebelli (see Leaf IV, No. 57), separates the cerebrum from the cerebellum. In the angles of the folds of the dura mater, blood-channels, called sinuses, are formed. (See Leaf IV, No. 58.) These drain the venous blood from the brain. The membranous coverings of the brain are called meninges.
4	Temporalis	(See Plate I, Leaf II, No. 2.)
5	Parietal Bones	(See Plate I, Leaf IV, Nos. 1 and 2.)
6	Occipital Bone	
7	Occipito frontalis	(See Leaf II, No. 20.)
8	Frontal Lobe of Cerebrum	(See Leaf IV, No. 1.) The cerebrum is concerned in the processes of thought, feeling, motion, and the special senses.
9	Temporo-sphenoidal Lobe of Cerebrum	(See Leaf IV, No. 3.)
10	Cerebellum	(See Leaf IV, No. 55.) The cerebellum is mostly concerned in co-ordination of motions.
12	Occipital Sinus	This drains into the torcular Herophili. (See Leaf IV, No. 58.)
13	Anterior Communicating Artery	The arterial blood supply of the brain is furnished by the internal carotid arteries (see Leaf II, No. 39) and the vertebral arteries (see Leaf III, No. 2) and their branches, as indicated in the plate. By the anterior communicating artery the two anterior cerebral arteries communicate. By the posterior communicating arteries the carotid arteries communicate with the posterior cerebral arteries, which are branches into which the basilar artery (see Leaf III, No. 2) divides. Thus a complete arterial circle, called the circle
14	Anterior Cerebral Arteries	
15	Middle Cerebral Arteries	
16	Internal Carotid Arteries	

17	Posterior Communicating Arteries	of Willis, surrounds the optic nerves, the pituitary body, and certain other important parts of the under surface of the cerebrum; from it are given off small vessels that supply all the interior parts of the cerebrum.
18	Posterior Cerebral Arteries	
19	Superior Cerebellar Arteries	
20	Anterior Cerebellar Arteries	
21	Basilar Artery	
22	Vertebral Arteries	
23	Anterior Spinal Artery	This arises by two roots from the vertebral arteries (see No. 22) which unite to form one vessel which descends along the front of the spinal cord.
24	Inferior Cerebellar Arteries	Branches from the vertebral arteries to supply the under parts of the cerebellum.
25	Pons Varolii	(See Leaf IV, No. 52.)
26	Medulla Oblongata	(See Leaf IV, No. 53.)
27	Pituitary Body	(See Leaf IV, No. 51.)
	Cranial Nerves	The nerves of special sense, those of the face, and those controlling certain organs that lie in the chest and abdomen, have their origin in the under parts of the brain and pass directly to the parts controlled. There are twelve pairs of such nerves, called the cranial nerves and designated by number, as follows:
i	Olfactory Nerves	These are large nerves with club-shaped ends which lie upon the upper surface of the ethmoid bone. From these ends nerve fibres descend into the upper parts of the nose, where are areas sensitive to odors. The nerves of smell. (See Plate IV, Fig. II, No. 14.)
ii	Optic Nerves	These are large nerves that partly cross each other and pass to the eyes. They are the nerves of sight. (See Plate IV, Fig. III.)
iii	Motor Oculi Nerves	These control the muscles that move the eyeball except two (see No. IV, and No. VI), and also the muscle fibres of the iris and the ciliary body. (See Plate IV, Fig. III.)
iv	Pathetici	These nerves supply the superior oblique muscles which turn the eyes downward and outward (a movement supposed to be the ocular method of expressing pathos). (See Plate IV, Fig. III, No. 55.)
v	Trigemini	These are the largest cranial nerves, each dividing into three branches, whence the name. A small branch of each of these nerves supplies the sense of taste to the front of the tongue. Another small branch controls the muscles of chewing. All the other branches supply sensation to the face, the teeth and gums, and the skin of the sides of the head.

vi	Abducentes	These nerves supply the external rectus, turning the eyeball outward. (See Plate IV, Fig. III, No. 28.)
vii	Facial Nerves	The nerves of motion, supplying the muscles of the face that have to do with expression. They run through bony canals so near the middle ear that often they are damaged by disease of the ear, and paralysis of the face results.
viii	Auditory Nerves	The nerves of hearing, passing to the internal ear. (See Plate IV, Fig. I, No. 24.)
ix	Glosso-pharyngeal Nerves	The nerves of sensation to the pharynx and tongue and a special nerve of taste to the back of the tongue.
x	Pneumogastric Nerves or Vagi	These supply the organs of voice and breathing with both motion and sensation, — and the pharynx, oesophagus, stomach, and heart with motion.
xi	Spinal Accessory Nerves	A part of each of these nerves joins the pneumogastric nerve to supply the pharynx and larynx, and the remainder supplies some of the neck muscles and helps to form the cervical plexus. (See Plate I, Leaf V, No. 11.)
xii	Hypoglossal Nerves	The motor nerves of the tongue.

Leaf VII. — THE UPPER SURFACE OF THE CEREBRUM

1	The Left Hemisphere	The great longitudinal fissure (3) divides the cerebrum into two hemispheres. Other fissures, which are constant, divide each hemisphere into lobes; and smaller fissures, which are more variable in their position and constancy, divide each lobe into convolutions. Certain portions of certain convolutions preside over definite actions or functions. The movements of the limbs are controlled in the upper part of the left parietal lobe; those of the head by an area in the upper part of the left frontal lobe. The face, speech, and hearing areas lie below the limb areas. The brain centre of sight is in the posterior portion of the left occipital lobe.
2	The Right Hemisphere	
3	The Great Longitudinal Fissure	
4	Frontal Lobe of the Cerebrum	
5	Occipital Lobe of the Cerebrum	
6	Parietal Lobe of the Cerebrum	

PLATE III
THE LARYNX, TOOTH, STOMACH, AND LIVER

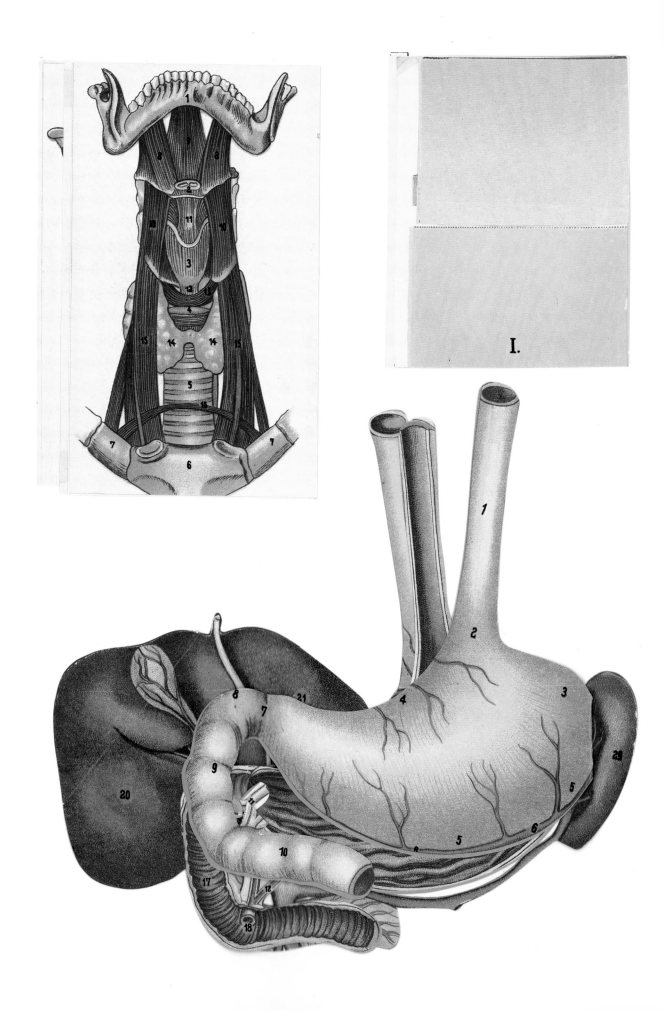

PLATE III

THE LARYNX, TOOTH, STOMACH, AND LIVER

FIGURE I. — THE UPPER RESPIRATORY ORGANS

The Larynx and its Supports

1	Inferior Maxillary Bone	Lower jawbone. (See Plate II, Leaf V, No. 9.)
2	Hyoid Bone	Shaped like the letter upsilon (the Greek letter u); hence its name. It is situated at the root of the tongue and supports it. It is suspended from the tip of the styloid process of the temporal bone. (See No. 62.) The muscles of the tongue are attached to it, and from it is suspended the larynx.
3	Thyroid Cartilage	(See No. 76.)
4	Cricoid Cartilage	(See No. 77 and No. 90.)
5	Trachea	(See Plate I, Leaf I, No. 1.)
6	Sternum	(See Plate I, Leaf III, Nos. 23–25.)
7	Cartilage of the First Rib	(See Plate I, Leaf III, No. 29.)
8	Hyoglossus	(See Plate II, Leaf II, No. 59.)
9	Mylo-hyoid Muscle	(See Plate II, Leaf II, No. 60.)
10	Thyro-hyoid Muscle	A flat band of muscle connecting the sides of the thyroid cartilage with the hyoid bone. It lifts the larynx and is practically a continuation of the sterno-thyroid muscle.
11	Thyro-hyoid Membrane	A tendinous sheet connecting the thyroid cartilage and the hyoid bone. It supports the larynx.
12	Crico-thyroid Membrane	This unites the cricoid cartilage with the thyroid cartilage, filling the space between them.
13	Crico-thyroid Muscle	This connects the cricoid and thyroid cartilages and controls their relative positions.
14	Thyroid Gland	(See Plate I, Leaf V, No. 52.)
15	Sterno-thyroid Muscle	This connects the thyroid cartilage with the sternum and is a depressor of the larynx.

	Longitudinal Sections through the Upper Respiratory Tract	
17	Superior Turbinated Body	
18	Middle Turbinated Body	(See Plate IV, Fig. II.)
19	Inferior Turbinated Body	
20	Superior Maxillary Bone	(See Plate I, Leaf III, No. 9.)
21	Anterior Palatine Canal	A canal passing downward through the bony roof of the mouth in the middle line. It is for the passage of vessels and nerves.
22	Mouth	
23	Tongue	
24	Soft Palate	(See Plate II, Leaf I, No. 16.)
25	Nasopharynx	That portion of the pharynx lying above the level of the soft palate. The nasal cavities open into it and the mouths of the Eustachian tubes are situated in it. This is an important region in its relations to disease of the nose and ears.
26	Tonsil	The tonsils are glandular masses situated on either side of the back portion of the mouth at its entrance into the pharynx. Upon their surface are openings into pockets, called crypts, which in their depths open into certain circulatory canals called lymphatics. The tonsils are prone to inflammations, in the course of which diseased material finds its way into the lymphatics and causes a serious disturbance of the health.
27	Anterior Pillar of the Tonsil	Membranous and muscular folds between which lies the tonsil.
28	Posterior Pillar of the Tonsil	
29	Upper Lip	
30	Lower Lip	
31	A Small Round Mirror, called a Laryngoscope, introduced into the back of the mouth at such an angle that in it may be seen the image of the interior of the Larynx and Trachea	
32	Inferior Maxillary Bone	(See Plate II, Leaf V, No. 9.)
33	Hyoid Bone	(See No. 2.)

34	Thyro-hyoid Membrane	(See No. 11.)
35	Thyroid Cartilage	(See No. 76.)
36	Epiglottis	A leaf-shaped flap of cartilage that shuts down over the top of the larynx in swallowing, thus keeping swallowed material from entering the larynx and trachea. It can be seen by depressing the back of the tongue. (See also Nos. 69 and 85.)
37	Cartilage of Wrisberg	(See No. 87.)
38	Arytenoid Cartilage	(See No. 89.)
39	False Vocal Cords	Fibrous bands covered with mucous membrane and lying above the true vocal cords. Their edges may be brought together, thus shutting the cavity of the larynx, as happens in explosive coughing. (See also No. 106.)
40	True Vocal Cords	Strong fibrous bands stretched from front to rear in the upper part of the larynx. The anterior ends lie near together, but the posterior ends are attached to movable cartilages, the arytenoids (see No. 89), and may be separated or brought together and made tense or relaxed, thus producing sounds of various pitch and intensity. (See No. 111.)
41	Ventricle of the Larynx	Upon each side of the larynx, between the true and the false vocal cords, is a recess called the ventricle of the larynx.
42	Cricoid Cartilage	(See No. 90.)
43	Trachea	(See Plate I, Leaf I, No. 1.)
44	Thyroid Gland	(See Plate I, Leaf V, No. 52.)
45	Oesophagus	(See Plate III, Fig. III.)

The Pharynx

46	Hard Palate	The palatal portion of the superior maxillary bone and the horizontal portion of the palate bone.
47	Inferior Maxillary Bone	(Sawn through.)
48	Superior Horns of the Thyroid Cartilage	(See No. 76.)
49	Thyroid Cartilage	(See No. 76.)
50	Posterior Nasal Spine	The spinous tip of the middle line of the hard palate.
51	Palato-pharyngeus	A thin sheet of muscle forming part of the soft palate and extending down the sides of the pharynx, forming the posterior pillar of the tonsil. (See No. 28.) It contracts the pharynx in swallowing.
52	Levator Palati	This forms part of the soft palate, arising from the roof of the pharynx. It lifts the palate against the back wall of the pharynx and so prevents food from entering the nasopharynx.

53	Uvula	A muscular and membranous appendage to the soft palate.
54	Superior Constrictor of the Pharynx	The muscular sheet forming the upper part of the bag-like cavity of the pharynx. By its contraction in swallowing, food is squeezed downward toward the oesophagus.
55	Middle and Inferior Constrictors of the Pharynx	These form the lower part of the pharynx and continue the action of the superior constrictor.

The Pharynx from behind

56	Occipital Bone	(See Plate I, Leaf IV, No. 1.)
57	Petrous Portion of the Temporal Bone	That part of the temporal bone which forms a portion of the floor of the cranium.
58	Eustachian Tube	(See Plate IV, Fig. I, No. 12.)
59	Septum of the Nose	(See Plate IV, Fig. II, Nos. 17–19.)
60	Turbinated Bodies	(See Plate IV, Fig. II, Nos. 5–7.)
61	Inferior Maxillary Bone	(See Plate II, Leaf V, No. 9.)
62	Styloid Process of the Temporal Bone	A strong, slender spine projecting downward from the under surface of the temporal bone, beneath the ear.
63	Tensor Palati	This extends from the sides of the soft palate to the under surface of the temporal bone. It stretches the soft palate.
64	Azygos Uvulae	A thin ribbon of muscle that raises the uvula. (See No. 53.)
65	Levator Palati	(See No. 52.)
66	Palatopharyngeus	(See No. 51.)
67	Uvula	(See No. 53.)
68	Tongue	
69	Epiglottis	(See No. 36.)
70	Pyriform Recess	A recess on either side of the base of the epiglottis.
71	Greater Horn of the Hyoid Bone	One of the extremities of the hyoid bone.
72	Larynx	
73	Trachea	(See Plate I, Leaf I, No. 1.)
74	Oesophagus	(See Fig. III, No. 1.)

The Larynx

75	Hyoid Bone	(See No. 2.)
76	Thyroid Cartilage	The largest and upper cartilage of the larynx. It consists of two wings, united in front at an acute angle, making

		the prominence of the neck called Adam's apple. The posterior edges of the wings, which are roughly square in shape, are extended upward into slender prolongations, called the superior horns of the thyroid cartilage (see Nos. 48, 84, and 86), and downward into short, stout prolongations called the inferior horns of the thyroid cartilage. (See No. 91.) To the inner surface of the acute angle are attached the anterior ends of the false and the true vocal cords and, above them, the epiglottis (see No. 36). By the tips of the lesser horns the thyroid cartilage is hinged upon the back part of the cricoid cartilage.
77	Cricoid Cartilage	This is shaped like a seal ring with the seal behind. To the sides of this posterior enlargement the lesser horns of the thyroid cartilage are attached, so that the thyroid cartilage tilts upon the cricoid cartilage. Upon the upper border of the seal-like portion of the cricoid cartilage are placed the arytenoid cartilages. (See 89.)
78	Trachea	(See Plate I, Leaf I, No. 1.)
79	Sterno-hyoid Muscle	(See No. 15.)
80	Thyro-hyoid Muscle	(See No. 10.)
81	Sterno-thyroid Muscle	(See No. 15.)
82	Thyro-hyoid Membrane	(See No. 11.)
83	Crico-thyroid Membrane	(See No. 12.)
84	Superior Horn of the Thyroid Cartilage	(See No. 76.)

Posterior View of the Larynx

85	Epiglottis	(See No. 36.)
86	Superior Horn of the Thyroid Cartilage	(See No. 76.)
87	Cartilages of Wrisberg	Small slender cartilages inserted into the fold of mucous membrane that extends from the tips of the arytenoid cartilages forward to the sides of the epiglottis, forming thus the upper border of the larynx.
88	Cartilages of Santorini	Two tiny cartilaginous bodies that are mounted upon the tips of the arytenoid cartilages.
89	Arytenoid Cartilages	Shaped like triangular pyramids, these small cartilages are mounted upon the posterior upper border of the seal-like portion of the cricoid cartilage. (See No. 77.) To their front surfaces are attached the posterior ends of the false vocal cords, near the summit, and of the true vocal cords, near the base. By the tilting of these cartilages and their rotation the cords are stretched or relaxed, and brought together or separated, in making sounds.
90	Cricoid Cartilage	(See No. 77.)

91	Inferior Horns of the Thyroid Cartilage	(See No. 76.)
92	Rings of the Trachea	As the figure shows, the rings are incomplete behind to allow a greater amount of freedom to the front of the oesophagus which lies against this surface of the trachea. (See Plate I, Leaf I, No. 1.)
93	Fibrous Portion of the Trachea	(See Plate I, Leaf I, No. 1.)
94	Membranous Covering of the Arytenoideus	The arytenoideus muscle connects the two arytenoid cartilages, drawing them together.
	Posterior View of the Nasal Cavities	
95	Roof of the Nasopharynx	
96	Turbinated Bodies	(See Plate IV, Fig. II, Nos. 5–7.)
97	Nasal Septum	(See Plate IV, Fig. II, Nos. 17–19.)
98	Mouth of the Eustachian Tubes	(See Plate IV, Fig. I, No. 12.)
99	Eustachian Eminence	The projecting posterior border of the mouth of the Eustachian tube, cartilaginous in structure.
100	Fossae of Rosenmüller	The upper, posterior angles of the nasopharynx. They are often filled with diseased glandular tissue which interferes with the proper action of the Eustachian tubes.
101	Soft Palate	(See Plate II, Leaf I, No. 16.)
102	Uvula	(See No. 53.)
	Views of the Larynx as Obtained by a Laryngoscope	(See No. 31.)
103	Epiglottis	(See No. 36.)
104	Pyriform Recess	(See No. 70.)
105	Rima Glottidis	The opening between the vocal cords.
106	False Vocal Cords	(See No. 39.)
107	True Vocal Cords	(See No. 40.)
108	Arytenoid Cartilages	(See No. 89.)
109	Cartilages of Wrisberg	(See No. 87.)
110	Cartilages of Santorini	(See No. 88.)
111	The Laryngoscopic Appearance of the Larynx during Phonation	During phonation, or the production of sound, the vocal cords are brought together to a variable extent and the current of air forced between them is set into vibration by them.

112	The Trachea and Bronchi as seen in the Laryngoscope	The cords are widely separated, as in breathing freely.

Figure II. — THE TOOTH

	Leaf I. — First Molar Tooth	(See Plate II, Leaf I, No. 19.)
1	Mucous Membrane of the Gum	The gum is composed of dense fibrous tissue closely adherent to the bone and covered with mucous membrane of limited sensibility.
2	Neck of the Tooth	The constricted portion of the tooth between the crown and the root.
3	Crown of the Tooth	That portion of the tooth that is above the gum.
4	Cusps	The rounded summits of the upper surface of the tooth. The canine teeth have but one cusp. The premolars or bicuspid teeth have two, and the molars have three, four, or five cusps.
5	Crucial Depression	The depression between the cusps.
	Leaf II. — The Roots and Sockets	
6	Gum	(See No. 1.)
7	Neck of the Tooth	(See No. 2.)
8	Roots, or Fangs	The roots are the portions of the teeth that sink into the depressions of the maxillary bone, where they are firmly embedded. The molar teeth have usually two or three roots, the other teeth but one. The roots of the canine teeth are very long.
9	Tips of the Roots	
10	Blood-Vessels	The nerves and vessels of the teeth enter at the tips of the roots.
11	Bone of the Jaw	
	Leaf III. — Section through the Tooth	
12	Mucous Membrane	The mucous membrane of the gum dips into the socket of the bone and is continuous with the periosteum lining of that cavity.
13	Fibrous Layer of the Gum	(See No. 1.)
14	Bone of the Jaw	
15	Enamel	The hardest part of the tooth, covering the surface of the crown. It is composed of hexagonal rods, resting on end upon the ivory. It is thickest upon the grinding surface, and disappears at the neck after gradually thinning.

16	Ivory or Dentine	This forms the body of the tooth. It is a modified bone tissue, its nutritive canals radiating from the hollow interior of the tooth.
17	Cement	A layer upon the surface of the root from the neck to the tip, and closely resembling bone in structure.
18	Dental Pulp	A soft and loosely formed substance which fills the hollow interior of the tooth. It is richly supplied with blood-vessels and nerves, which enter at the tips of the roots. It is very sensitive.
19	Pulp Cavity of the Root	
20	Vessels and Nerves	

FIGURE III. — THE ORGANS OF DIGESTION

1	Oesophagus	A muscular walled tube leading from the pharynx to the stomach, lying behind the trachea and joined to it. (See Fig. I, No. 92.) By a stripping action of its muscles it forces swallowed food downward into the stomach.
2	Cardiac End of the Stomach	The end nearer the heart.
3	Fundus of the Stomach	The large end of the stomach, lying upon the left side just under the diaphragm and near the front of the abdomen.
4	Lesser Curvature of the Stomach	The upper surface of the stomach.
5	Greater Curvature of the Stomach	The lower surface of the stomach.
6	Gastro-epiploic Artery	The artery of the greater curvature, made from branches of the hepatic (liver) artery and the splenic (spleen) artery, these two being branches from one trunk. (See No. 13.)
7	Pylorus	The end of the stomach opening into the intestine. The name means "a portal" or "opening."
8	Ascending Portion of the Duodenum	The duodenum is the first portion of the small intestine, its length being equal to the breadth of twelve fingers, which gives the name. The ducts from the liver and pancreas enter below the middle of the descending portion.
9	Descending Portion of the Duodenum	
10	Transverse Portion of the Duodenum	
11	Common Bile Duct	The duct of the liver and the gall bladder.
12	Pancreatic Duct	The duct of the pancreas (See No. 34.)
13	The Posterior Surface of the Stomach	This diagram shows the distribution of the gastric artery, which, with the hepatic artery and the splenic artery (see No. 6), springs from one trunk, the coeliac axis, which is given off from the abdominal aorta. (See Plate I, Leaf I, No. 22.)

14	Inner Surface of the Stomach	A thick, velvety, mucous membrane, thrown into numerous folds, called rugae, and richly supplied with the glands that secrete the gastric juice.
15	Section of the Wall of the Stomach	The wall of the stomach consists of a thin serous membrane on the outside, a muscular coat beneath the serous coat, and mucous membrane lining the inner surface. The fibres of the muscular coat run lengthwise, obliquely and around the organ.
16	Pyloric Valve	The opening into the duodenum. Here the circular muscular fibres of the stomach wall form a dense muscular ring which closes the opening.
17	Inner Surface of the Duodenum	Here are shown circular folds of the mucous membrane, the valvulae conniventes, by which a greater surface is presented for the absorption of nutriment from the digested contents of the intestine.
18	Orifice of the Bile and Pancreatic Ducts	(See Nos. 11 and 12.)
19	Posterior Surface of the Stomach (*The Liver is represented as tipped up and back.*)	Showing the muscular fibres. (See No. 15.)
20	The Right Lobe of the Liver	The liver is the largest gland in the body, and is situated under the right side of the diaphragm. It secretes bile and also effects important changes in the blood that passes through it. It consists of five lobes, separated from one another by fissures — the largest lobe being the right lobe. The left lobe is next in size, and the other three are very small lobes on the under surface. Upon this surface is the deep transverse fissure into which enter the large vessels of the liver and out of which proceeds the hepatic duct, which joins with the cystic duct from the gall bladder to make the common bile duct (No. 11). The portal vein (vena portae) brings to the liver blood from the intestine. Within the liver this vessel divides into very small branches (portal vessels) which are in close proximity to a second set of tiny vessels (hepatic vessels) that run together and form eventually the hepatic vein. This vein takes to the inferior vena cava the blood which has been filtered in its passage through the liver from the portal vessels into the hepatic vessels. (See Plate I, Leaf I, Nos. 15 and 16.) The gall bladder is a storage reservoir for bile.
21	The Left Lobe of the Liver	
22	Gall Bladder	
23	Transverse Fissure of the Liver	
24 to 25	Vena Portae	
26	Splenic Vein	The vein from the spleen.
27	Coronary Veins	The veins of the lesser curvature, opening into the vena portae (No. 24).
28	Right Gastro-epiploic Vein	The vein of the greater curvature, opening into the superior mesenteric vein, which is tributary to the vena portae (No. 24).

29	Spleen	A ductless gland, situated at the right of the stomach just under cover of the lower ribs. It has to do with blood elaboration.
30	Upper Surface of the Liver	A smooth dome-shaped surface of the right and left lobes, in contact with the diaphragm.
31	Posterior Surface of the Stomach	
32	Anterior Surface of the Pancreas	A long gland, shaped like a dog's tongue and situated behind the stomach. Its large curved head lies in the angle between the descending and transverse portions of the duodenum, and its long body lies horizontally, extending by its tail to the spleen. Its duct opens into the duodenum with the common bile duct (No. 12).
33	Posterior Surface of the Pancreas	
34	Pancréatic Duct	

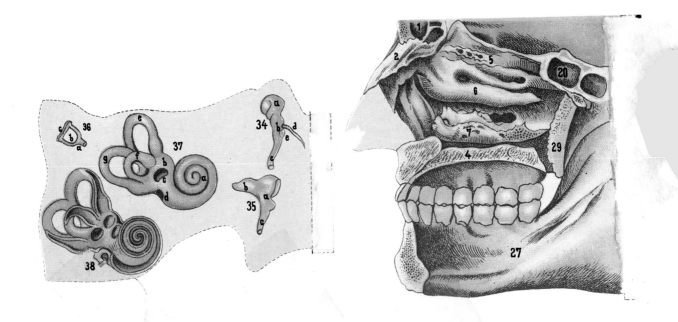

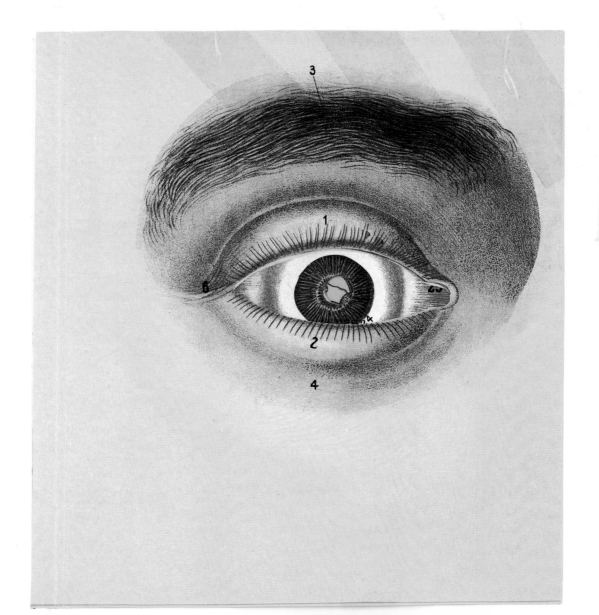

PLATE IV

THE EAR, EYE, AND NOSE

PLATE IV

THE EAR, EYE, AND NOSE

Figure I. — THE EAR

1	Auricle, or Pinna	A cartilaginous, shell-like expansion of the external ear.
2	External Meatus (Cartilaginous Portion)	The pinna and the meatus, or canal, of the ear, make up the external ear.
3	External Meatus (Bony Portion)	
4	Temporal Bone	(See Plate I, Leaf III, No. 3, and Plate III, Fig. I, No. 57.) The structures of the ear lie in the petrous portion of the temporal bone.
5	Styloid Process of the Temporal Bone	(See Plate III, Fig. I, No. 62.)
6	Membrana Tympani	The ear-drum membrane. (See No. 16.)
7	Posterior Semicircular Canal	
8	Superior Semicircular Canal	(See No. 22.)
9	External Semicircular Canal	
10	Fenestra Ovalis	The oval window. (See No. 18.)
11	Cochlea	"The snail." (See No. 23.)
12	Eustachian Tube	(See No. 21.)
13	Internal Carotid Artery	This enters the cranium by a very crooked canal through the petrous portion of the temporal bone. (See Plate II, Leaf II, No. 39, and Leaf VI, No. 16.)

Schematic Section through the Ear

14	Auricle	(See Nos. 1–3.)
15	External Meatus	
16	Membrana Tympani	The middle ear, or tympanum (drum), is a cavity in the petrous portion of the temporal bone. It is separated from the external ear by a tense and elastic membrane — the membrana tympani — and from the internal ear by a bony wall that is pierced by two openings, — the fenestra ovalis, or oval window, and the fenestra rotunda, or round window, both of these being closed by a thin membrane. The middle ear communicates with the air by means of the Eustachian tube which opens from it into the nasopharynx. (See Plate II, Leaf IV, No. 20.) The middle ear also commu-
17	Tympanum	
18	Fenestra Ovalis	
19	Fenestra Rotunda	
20	Ossicles	
21	Eustachian Tube	

nicates with the system of cells in the mastoid bone. (See Plate II, Leaf V, No. 19.) In the middle ear is suspended the chain of ossicles that extends from the drum membrane to the oval window (see Nos. 34–36) and by and through this chain vibrations of air are transmitted from the drum membrane to the contents of the internal ear where the nerve of hearing is situated.

The internal ear, or labyrinth as it is called on account of the complexity of its shape, is situated in the petrous portion of the temporal bone, internal to the middle ear. The round and oval windows of the middle ear open into a chamber called the vestibule. Anteriorly the vestibule opens into a spiral chamber like that of a snail shell, called on this account the cochlea. This spiral chamber is made three-barrelled in the following manner : A spiral sheet of bone, called the lamina spiralis, springs from the wall of the cochlea nearest its axis and reaches half-way to the outer wall. From its edge a fibrous membrane, the basilar membrane, stretches to the outer wall, dividing the chamber into nearly equal upper and lower chambers. From the edge of the lamina spiralis a second membranous sheet reaches to the outer wall, making an acute angle with the basilar membrane and encroaching upon the upper of the two divisions completed by that membrane. This is called the membrane of Reissner. Thus are formed three spiral chambers ; that below the basilar membrane is called the scala tympani, the upper one is called the scala vestibuli, and the triangular one between the basilar membrane and Reissner's membrane is called the scala media. Along the upper surface of the basilar membrane near its attachment to the lamina spiralis lies a complex structure of nerve and membranous tissue called the organ of Corti. It is the terminal organ of the nerve of hearing. (See Plate II, Leaf VI, No. VIII.) Posteriorly the vestibule opens into three circular canals whose planes are at right angles with each other. These are the semicircular canals. The membranous lining of the cochlea is continuous with that of the vestibule and the semicircular canals, but in these cavities the membrane lies loose from the bony walls and is formed into two sacs in the vestibule, — the saccule, which opens into the cochlea, and the utricle, which opens into the membranous semicircular canals, each semicircular canal having a dilatation at its beginning, called its ampulla. The nerve of hearing enters the posterior surface of the temporal bone and divides into two parts, — one being distributed to the organ of Corti and the other to the vestibule.

22	Semicircular Canal
23	Cochlea
24	Auditory Nerve
25	Vestibular Branch of the Auditory Nerve
26	Cochlear Branch of the Auditory Nerve
27	Organ of Corti
28	Scala Vestibuli
29	Scala Tympani
30	Utricle
31	Saccule
32	Membranous Semicircular Canal
33	Ampulla

The Ossicles

34	The Malleus
a	Head
b	Neck
c	Manubrium or Handle
d	Processus Gracilis

The hammer. Its head lies high in the tympanum and its manubrium lies against and is set into the fibres of the drum membrane, the tip of the manubrium marking the centre of the drum membrane. The processus gracilis extends forward along the outer wall of the tympanum and the short process is set into the upper edge of the drum

e	Short Process	membrane. The head is closely joined to the body of the incus by a movable joint.
35	Incus	The anvil. This lies above and internal to the malleus and somewhat posterior to it. By its notched surface the body is closely joined to the head of the malleus and is bound to it by ligaments. The short process extends backward and rests in a notch in the posterior part of the tympanum. The long process reaches down and back in the tympanum to the level of the oval window, where its tip is joined to the head of the stapes.
a	Body	
b	Short Process	
c	Long Process	
36	Stapes	The stirrup. Its head is joined to the tip of the long process of the incus and its footplate is set into the oval window, being adherent to the membrane that closes that window.
a	Head	
b	Crura (Legs)	
c	Footplate	
		By the arrangement of the ossicles vibrations of the drum membrane are passed through the chain of bones and communicated to the contents of the internal ear, there affecting the terminals of the nerve of hearing.
37	Cast of the Internal Ear	This shows the relations of the parts of the internal ear.
a	Cochlea	
b	Vestibule	
c	Oval Window	
d	Round Window	
e	Superior Semicircular Canal	The semicircular canals are supposed to control the balance of the body. Injury to them, or disease of them, causes dizziness.
f	External Semicircular Canal	
g	Posterior Semicircular Canal	
38	The Bony Labyrinth (laid open)	The internal ear cavities are called the labyrinth on account of their complexities.

FIGURE II. — THE NOSE

1	Frontal Sinus	A cavity lying in the angle of the frontal bone, above the orbit and opening downward into the nose.
2	Nasal Bones	These form the bridge of the nose.
3	Upper Lip	
4	Hard Palate	(See Plate III, Fig. I, No. 46.)
5	Superior Turbinated Body	Shell-like masses protruding into the nasal cavity, the superior one very small, the middle turbinated body of variable size and usually hollow, and the inferior body the largest and longest one. They are covered with mucous membrane which is richly supplied with mucous glands and blood. In breathing, the air passes over and between these bodies, in air-passages, and by them is warmed and
6	Middle Turbinated Body	
7	Inferior Turbinated Body	

8	Mouth	made moist. (See Plate III, Fig. I, No. 96.) Each of these air-passages is called a meatus. (See Nos. 21–23.)
9	Nasopharynx	That part of the pharynx into which the nasal cavities open. (See Plate III, Fig. I, No. 25.)
10	Epiglottis	(See Plate III, Fig. I, No. 36.)
11	Tongue	
12	Lower Lip	
13	Inferior Maxillary Bone	
14	Olfactory Nerve	The terminals of the nerve of smell (see Plate II, Leaf VI, No. I), situated in the upper portion of the septum (see No. 17) and the surface of the superior turbinated body. (See No. 5.)
15	Nasal Nerve	A branch of the superior division of the fifth cranial nerve. (See Plate II, Leaf VI, No. V.)
16	Nasopalatine Nerve	A branch of the middle division of the fifth cranial nerve. (See Plate II, Leaf VI, No. V.)

(*The Mucous Membrane is pictured as removed from the Septum in Nos. 17, 18, and 19.*)

17	Perpendicular Plate of the Ethmoid Bone	These three make up the septum, which is the partition between the two nasal cavities. Its anterior portion is of cartilage and the posterior portion of bone.
18	Vomer (bone)	
19	Triangular Cartilage	
20	Sphenoidal Sinus	A cavity in the body of the sphenoid bone. (See Plate II, Leaf IV, Nos. 15–16.) Its opening lies high in the posterior wall of the nasal chamber and it is a common source of nasal disease.
21	Superior Meatus	(See No. 5.) The sense of smell has its sensory area above and in the superior meatus. The middle meatus is the principal breathing passage, and the lower meatus is of chief use for drainage. Into the lower meatus opens the nasal duct from the eye. (See Fig. III, No. 18.) In the upper part of the nose are many cavities which are cells in the ethmoid bone. (See Plate II, Leaf IV, No. 14.) They open and drain into the middle and superior meatus and are often the seat of nasal disease. Into the middle meatus also opens the large cavity in the superior maxillary bone, called the antrum of Highmore. This also is sometimes the seat of nasal disease.
22	Middle Meatus	
23	Inferior Meatus	
24	Upper Lateral Cartilage	These form the alae, or wings, of the nose.
25	Lower Lateral Cartilage	
26	Superior Maxillary Bone	(See Plate I, Leaf III, No. 9.)
27	Inferior Maxillary Bone	(See Plate II, Leaf V, No. 9.)
28	Orbit of the Eye	(See Fig. III, Nos. 58–61.)
29	Wing of the Sphenoid Bone (in section)	(See Plate II, Leaf IV, No. 15.)

Figure III. — THE EYE

1	Upper Lid	(See Plate II, Leaf II, Nos. 2–3.)
2	Lower Lid	
3	Eyebrow	
4	Lower Border of the Orbit	
5	Inner Canthus, or Angle, of the Eye	
6	Outer Canthus, or Angle, of the Eye	
7	Meibomian Glands	These glands are situated upon the inner surface of the eyelids and open by ducts upon the free margin of the lids through minute mouths. They secrete fatty material.
8	Orifices of the Meibomian Glands	
9 to 10	Tarsal Plates	Cartilaginous plates, having thick straight edges, set into the lids and furnishing the even margins of the lids that come into close contact upon closing the eyes.
11 to 12	Internal and External Tarsal Ligaments	These support the eyelids and the tarsal plates.
13	Superior and Inferior Tarsal Ligaments	These support the tarsal plates.
14	Lachrymal Caruncle	A small, reddish, glandular structure at the inner canthus of the eye.
15a 15b	Palpebral Portion of Orbicularis Palpebrarum	(See Plate II, Leaf II, No. 3.)
16 17 18 19 20	Puncta Lachrymalia, Lachrymal Ducts, Lachrymal Sac, Nasal Duct	Puncta lachrymalia are openings upon the margins of the lids into the lachrymal ducts. These ducts are short canals that converge and meet in the lachrymal sac, which is an enlargement of the upper end of the duct (nasal duct) that leads downward into the nose, opening into the lower meatus. (See Fig. II, No. 23.) Excessive tear supply is drained from the eye by these channels.
20a	Lachrymal Glands	These are situated in a recess of the outer roof of the orbit just inside the border. (See No. 71.)
20b	Openings of Ducts from the Lachrymal Glands	
21 to 24	Sclerotic Coat of the Eye, or Sclera	A dense, hard, firm, spherical shell that maintains the form of the eyeball. Through its back wall is a small opening for the entrance of the optic nerve. It is thicker behind, and in front is continuous with the cornea. (See No. 29.)
25	Internal Rectus Muscle	This rotates the eyeball toward the nose.
26	Inferior Rectus Muscle	This rotates the eyeball downward.
27	External Rectus Muscle	This rotates the eyeball away from the nose.

28	Superior Rectus Muscle	This rotates the eyeball upward.
29	Cornea (mica)	The projecting transparent part of the outer coat of the eye, forming the anterior sixth of the globe. It is a continuation of the sclera. (See No. 21.)
30	Choroid Coat	A thin lining of the inner surface of the sclera, rich in blood-vessels and continuous with the iris (see Nos. 32–34), about whose circumference the choroid is thickened into the ciliary processes. (See No. 37.)
31	Choroid Blood-Vessels	
32 to 34	Iris	The colored portion of the front of the eye. It is a muscular disc with a round hole, the pupil (see No. 35), in the centre. The muscular fibres are both circular and radiating, the one set diminishing the size of the pupil and the other enlarging it.
35	Pupil	
36	Ora Serrata	The anterior border of the retina. (See No. 40.)
37	Ciliary Processes	A thickened ring of choroid tissue about the iris (see No. 30) mingling with fibres of the ciliary muscle which regulates the shape of the crystalline lens. (See No. 38.)
38	Crystalline Lens (mica)	A transparent, double-convex lens situated behind the iris. The amount of its convexity may be to a certain extent regulated by the ciliary muscle, thus focussing rays of light from near or far objects upon the retina.
39	Vitreous Humor (mica)	A transparent substance, of the consistency of thin jelly, that fills the concavity of the eyeball behind the crystalline lens.
40	Retina	The retina is the inner coat of the eye, the one sensitive to light. It is a membrane of complex structure and covers the interior of the eyeball as far forward as the iris, where it ends in the ora serrata. (See No. 36.) Its most sensitive area is exactly opposite the pupil in a slightly elevated yellow spot, the macula lutea (see No. 43), with a central depression, the forea centralis. (See No. 43.) The optic nerve enters the eyeball just internal to the macula and spreads out in the retina. The artery of the retina enters through the centre of the optic nerve and spreads out in the retina.
41	Optic Nerve	
42	Vessels of the Retina	
43	Macula Lutea and Forea Centralis	
44 47 48	Blood-Vessels of the Choroid	(See No. 30.)
45	Blood-Vessels of the Optic Nerve	(See No. 41.)
46	Pigment Layer of the Retina	The innermost (surface) layer of the retina.
49	Veins of the Choroid	A star-like arrangement of veins called venae vorticosae.
50	Sclera	(See No. 21.)
51	Optic Nerve	(See Plate II, Leaf VI, No. II.)
52	Ciliary Arteries	These converge toward the ciliary body and supply the iris.

53	Outer Layer of the Choroid	
54	Ciliary Nerves	These supply motion to the ciliary muscles. (See No. 37.)
55	Superior Oblique Muscle	This muscle passes through a fibrous loop (the Pulley) in the upper inner portion of the roof of the orbit, near its anterior margin (No. 72), and turning outward is attached to the upper surface of the eyeball. It rotates the eyeball downwards and away from the nose.
56	Inferior Oblique Muscle	Arising from the inner wall of the orbit near the lachrymal duct, this muscle runs under the eyeball and outward and backward, being attached to the outer surface of the eyeball. It rotates the eye upward and outward.
57 57a	Optic Nerve	This nerve enters the orbit through the optic foramen (see No. 73), and runs forward to enter the eyeball just inside its antero-posterior axis. (See No. 40.)
58	Orbital Surface of the Frontal Bone	
59	Orbital Surface of the Sphenoid Bone	
60	Orbital Surface of the Malar Bone	(See Plate I, Leaf III, No. 8.)
61	Orbital Surface of the Superior Maxillary Bone	
62	Spheno-maxillary Fissure	This fissure is at the junction of the outer wall and floor of the orbit, transmitting a branch of the fifth cranial nerve (see Plate II, Leaf VI, No. V) and some infraorbital blood-vessels.
63	Lachrymal Bone	This bone forms part of the inner wall of the orbit. The nasal duct lies in a groove upon its surface. (See No. 70.)
64	Ethmoid Bone	(See Plate II, Leaf IV, No. 14.)
65	Orbital Border of the Frontal Bone	
66	Orbital Border of the Superior Maxillary Bone	
67	Supraorbital Foramen	For the exit of the supraorbital nerve. (See Plate II, Leaf III, No. 15.)
68	Infraorbital Foramen	The termination of the infraorbital canal, a groove upon the floor of the orbit, transmitting the infraorbital blood-vessels and nerve, the former being branches of the internal maxillary artery, and the latter a branch of the middle division of the fifth cranial nerve. (See Plate II, Leaf VI, No. V.)
69	Ethmoidal Foramina	Openings through the inner wall of the orbit transmitting the ethmoidal vessels, branches of the ophthalmic division of the internal carotid artery, and the nasal nerve which is a part of the first division of the fifth cranial nerve. (See Plate II, Leaf VI, No. V.)

4

70	Lachrymal Groove	This is for the nasal duct. (See No. 20.)
71	Lachrymal Fossa (*Ditch*)	This lies in the upper outer angle of the orbit for the lachrymal gland. (See No. 20, a.)
72	Attachment of the Pulley for the Superior Oblique Muscle	(See No. 55.)
73	Optic Foramen	The opening into the orbit for the entrance of the optic nerve. (See Plate II, Leaf VI, No. II.)
74	Sphenoidal Fissure	A long slit between the greater and lesser wings of the sphenoidal bone, situated in the roof of the orbit, for the transmission of the third, the fourth, the three branches of the first division of the fifth and the sixth cranial nerves (see Plate II, Leaf VI, Nos. III–VI), a branch from the middle meningeal artery (see Plate II, Leaf IV, No. 9), and the ophthalmic vein, with some fibres from a sympathetic nerve plexus.
75	Spheno-maxillary Fissure	(See No. 62.)
76	Origin of the Internal Rectus Muscle	
77	Origin of the Inferior Rectus Muscle	(See Nos. 25–28.)
78	Origin of the External Rectus Muscle	NOTE. — The more fixed attachment of a muscle (the point from which it pulls) is called its origin; the attachment at the point where it causes motion is called its insertion.
79	Origin of the Superior Rectus Muscle	
80	Origin of the Levator Palpebrae	This lifts the eyelid.
81	Origin of the Superior Oblique Muscle	(See No. 55.)